Carbon Isotope Techniques

Isotopic Techniques in Plant, Soil, and Aquatic Biology

Series Editors

Eldor A. Paul
Department of Crop and Soil Science
Michigan State University
East Lansing, Michigan

Jerry M. Melillo
The Ecosystems Center
Marine Biological Laboratory
Woods Hole, Massachusetts

Carbon Isotope Techniques

Review Board

C. V. Cole
Natural Resource Ecology Laboratory
Colorado State University
Fort Collins, Colorado

D. A. Crossley, Jr.
Institute of Ecology
University of Georgia
Athens, Georgia

Carbon Isotope Techniques

Edited by

David C. Coleman
Department of Entomology
University of Georgia
Athens, Georgia

Brian Fry
The Ecosystems Center
Marine Biological Laboratory
Woods Hole, Massachusetts

ACADEMIC PRESS, INC.
Harcourt Brace Jovanovich, Publishers
San Diego New York Boston
London Sydney Tokyo Toronto

Academic Press, Inc.
San Diego, California 92101

United Kingdom Edition published by
Academic Press Limited
24–28 Oval Road, London NW1 7DX

Library of Congress Cataloging-in-Publication Data

Carbon isotope techniques / edited by David C. Coleman, Brian Fry.
       p.   cm. -- (Isotopic techniques in plant, soil, and aquatic
   biology series)
       Includes index.
       ISBN   0-12-179730-9   (hardcover)(alk. paper)
       ISBN   0-12-179731-7   (paperback)(alk. paper)
       1. Stable isotope tracers.   2. Biology--Technique.   3. Carbon-
   -Isotopes.   I. Coleman, David C., date.   II. Fry, Brian.
   III. Series.
   QH324.3.C37   1991
   574.19'285--dc20   90-25392
                                                           CIP

PRINTED IN THE UNITED STATES OF AMERICA
91   92   93   94   9   8   7   6   5   4   3   2   1

Contents

9 *Bomb Carbon*
K. M. Goh

II USES AND PROCEDURES FOR ^{13}C

10 *Stable Carbon Isotope Ratios of Natural Materials:*
I. Sample Preparation and Mass Spectrometric Analysis
Thomas W. Boutton

11 *Stable Carbon Isotope Ratios of Natural Materials:*
II. Atmospheric, Terrestrial, Marine, and
Freshwater Environments
Thomas W. Boutton

Contributors

Numbers in parentheses indicate the pages on which the authors' contributions begin.

L. A. Bjelk (101), Department of Horticultural Science, North Carolina State University, Raleigh, North Carolina 27695

T. W. Boutton (155, 173, 219), Department of Rangeland Ecology and Management, Texas Agricultural Experiment Station, Texas A&M University, College Station, Texas 77843

D. C. Coleman (3), Department of Entomology, University of Georgia, Athens, Georgia 30602

F. T. Corbin (3, 101), Department of Crop Science, North Carolina State University, Raleigh, North Carolina 27695-7627

J. R. Ehleringer (187), Department of Biology, University of Utah, Salt Lake City, Utah 84112

J. N. Gearing (201), Maurice – Lamontagne Institut, Fisheries and Oceans Canada, Mont-Joli, Quebec G5H 3Z4, Canada

K. M. Goh (125, 147), Department of Soil Science, Lincoln College, Canterbury, New Zealand

E. G. Gregorich (77), Land Resource Research Centre, Agriculture Canada, Central Experimental Farm, Ottawa, Ontario K1A 0C6, Canada

D. Harris (39), Department of Crop and Soil Science, Michigan State University, East Lansing, Michigan 48824

J. Kummerow (11), Department of Biology, College of Sciences, San Diego State University, San Diego, California 92182-0057

A. E. McElroy (109), Environmental Sciences Program, University of Massachusetts at Boston, Harbor Campus, Boston, Massachusetts 02125

T. J. Monaco (101), Department of Horticultural Science, North Carolina State University, Raleigh, North Carolina 27695

E. A. Paul (39), Department of Crop and Soil Science, Michigan State University, East Lansing, Michigan 48824

A. M. Pregnall (53), Biology Department, Vassar College, Poughkeepsie, New York 12601

P. J. H. Sharpe (245), Biosystems Research Division, Department of Industrial Engineering, Texas A&M University, College Station, Texas 77843

R. D. Spence (245), Division of Science and Engineering, University of Texas of the Permian Basin, Odessa, Texas 79762

R. P. Voroney (77), University of Guelph, Ontario Agricultural College, Department of Land Resource Science, Guelph, Ontario N1G 2W1, Canada

F. R. Warembourg (11), Centre d'Ecologie Fonctionnelle et Evolutive, Centre Nationale de la Reserche Scientifique, CNRS Centre L. Emberger, F-34033 Montpellier Cedex, France

J. P. Winter (77), University of Guelph, Ontario Agricultural College, Department of Land Resource Science, Guelph, Ontario N1G 2W1, Canada

Preface

Historically, isotopes have played an important role as diagnostic tools in natural or human-modified experiments throughout a wide range of biological studies. As various disciplines have matured and instrumentation has become more sophisticated and readily available, a range of opportunities for use of stable isotopes has been added to the impressive repertoire of radioisotopes that have been in general use since the 1960s.

When we were asked to edit a book on carbon isotopes in plant, soil, and aquatic biology, our response was enthusiastic. We felt there was a real need for a user-oriented book that would be of interest to a wide audience. Our authors have written explicitly for the advanced undergraduate or graduate student, as well as any well-rounded generalist scientist who has not previously used radioisotopes or stable isotopes in his or her research. This book is meant for frequent use; its most suitable habitat is a laboratory bench or laboratory desk. It is designed for easy perusal as people plan laboratory or field research.

Perhaps at no time in the history of biology has there been such a profusion of new techniques and tools developed for analytical purposes. This is most evident at the subcellular, cellular, tissue, and organ levels, since a wide variety of molecular genetic and biochemical techniques have come into use.

Of perhaps equal intensity, but less evident in popular science publications (e.g., *Science, Scientific American*), has been a merger of research objectives in the disciplines of physiological ecology and ecosystems studies. The main impetus has come from the revelation that "warm season" plants with the ^4C pathway (or Hatch-Slack pathway, with ^4C compounds as principal components of intermediary metabolism) have a significantly different content of ^{13}C than do "cool season" plants with the ^3C (or Calvin-Benson) pathway. Implications of this basic separation on a physiological basis are extended and amplified through entire ecosystems, from plants into soils, and in aquatic systems, as noted in several of the chapters in our book (see especially chapters by Boutton and Ehleringer).

For finer levels of resolution, particularly necessary when initial starting materials are considerably diluted (e.g., see Chapter 5 by Voroney *et al.* and Chapters 8 and 9 by Goh on long-term soil studies, or see Chapter 6 by Corbin *et al.* and Chapter 7 by McElroy on aspects of herbicides and environmental toxicology) it is advisable to measure constituents labeled with the radioisotope ^{14}C. This isotope also has been introduced by humankind via nuclear weapons tests (cf. Goh) and by cosmic rays, and is traceable with a variety of techniques.

What we, the editors, find both challenging and exciting is that the approaches contained herein enable workers from diverse backgrounds to bring their analytical tools to bear on a range of whole-system problems. For example, who would have suspected that physiological ecologists, experimenting on plant translocation of photosynthates, would have anything in common with air pollution researchers? Thus short-term changes in key plant physiological parameters (such as stomatal conductance, carbon exchange rates, and phloem transport of carbon in shoots and roots) can be assessed using the short-lived gamma-emitting isotope ^{11}C (half-life $= 20.3$ minutes), as noted by Spence and Sharpe.

We welcome any readers' suggestions and comments about additional areas of interest, or possible changes and additions to existing chapters and protocols.

David C. Coleman
Brian Fry

I

Uses and Procedures for ^{14}C

1

Introduction and Ordinary Counting as Currently Used

D. C. Coleman

Department of Entomology
University of Georgia
Athens, Georgia 30602

F. T. Corbin

Department of Crop Science
North Carolina State University
Raleigh, North Carolina 27695

I. INTRODUCTION

A. Units of Measure

Natural radioactivity cannot be increased or decreased by any ordinary process. The activity of a quantity of radioactive material is the number of nuclear disintegrations that occur in unit time. The unit of activity was described originally as the curie and was defined as the radioactivity in one gram of radium. The value of the curie thus was dependent upon experimental measurement and was finally stabilized at 3.7×10^{10} disintegrations per second (DPS).

The SI (Systéme Internationale) unit of radioactivity, the Becquerel (Bq), is the preferred unit for current publication in professional journals. Also the unit value of the Becquerel (1 Bq = 1 DPS) is the desired number to follow in calculations, in preference to the microcurie (1 μCi = 37,000 DPS). Some of the more common terms in current use on product labels and in technical literature and the appropriate conversion factors are listed in Table 1.

B. Specific Activity

An accurate analysis of the amount of radioactive substances in soils and plants requires that the radiolabel be described as a function of concentration. *Specific activity* is the rate of decay per unit mass of an element and is usually listed as mCi/mmol, or kBq/mmol. The maximum specific activity, in which every molecule of a substance contains the radiolabel, is seldom

Table 1
Units of Radioactivity and Corresponding DPM Values

Unit	Fraction of unit	DPS (disintegrations per second)	DPM (disintegrations per minute)
Curie (Ci)	10^0	3.7×10^{10}	2.22×10^{12}
Milli-Curie (mCi)	10^{-3}	3.7×10^7	2.22×10^9
Micro-Curie (μCi)	10^{-6}	3.7×10^4	2.22×10^6
Nano-Curie (nCi)	10^{-9}	3.7×10^1	2.22×10^3
Becquerel (Bq)	10^0	1	60
Kilo-Becquerel (kBq)	10^3	1×10^3	60×10^3
Mega-Becquerel (MBq)	10^6	1×10^6	60×10^6

Note: SI notation uses Becquerels exclusively.

attained in actual practice; usually only one or two radioactive atoms per 10^6 atoms total are needed in radiotracer work. The constant, λ, is known as the *decay constant* and is expressed in units of time. The rate of decay is proportional to the number of radioactive atoms present.

1. Example Calculation

Carbon (^{14}C) is continually produced in the upper atmosphere by cosmic radiation and is relatively constant in nature. What is the maximum specific activity of $K_2^{14}CO_3$? One mole contains 6.023×10^{23} radioactive carbon atoms (Avogadro's number).

$$\text{Specific activity} = \lambda N$$

$$\text{Specific activity} = \frac{(0.00012 \text{ yr})(6.023 \times 10^{23})}{(365 \text{ days})(24 \text{ hr})(60 \text{ min})(60 \text{ sec})}$$

$$\text{Specific activity} = 62 \text{ Ci/mol}$$

where λ is the *decay constant* and N is the number of radioactive carbon atoms in 1 mole.

C. Radiation Detection

There are several radiation-counting instruments that can be employed for measuring the radioactivity in samples containing carbon-14.

1. Gas Counters

Particle counters for measurement of radioactive emissions were designed by Rutherford and Geiger and later extensively developed by Geiger and Müller. Similar counters are in use today and the current sophisticated microprocessor-controlled Imaging Proportional Counters and thin layer chromatography (TLC) scanners operate on the basic principles of the early instruments. Operation is based on the phenomenon that large numbers of ions are formed during the passage of charged particles, such as alpha and beta emissions, through a gas. The number of ions produced by a beta particle is not very large, but when a high voltage (1000 volts) is connected in series with the counter, a cascade effect results, which amplifies the original current, and the resultant current can be detected with ease and quantified.

2. Scintillation Counters

Experiments by Becquerel on the fluorescence of substances during exposure to X rays resulted in one of the early ways of detecting nuclear particles. Flashes of light, or scintillations, occur in certain crystals, such as naphthalene and anthracene, when they are exposed to nuclear particles. Light is emitted with frequencies characteristic of the atoms of the crystal, but this process had only limited use for many years because the light intensities of each scintillation were too low for accurate measurements. In recent years, the photomultiplier tube and associated circuitry have served to multiply the photoelectric currents from scintillators by a large factor. These tubes can be incorporated into amplifying and counting circuits for counting particles up to a million per second.

3. Liquid Scintillation Counting

The most frequently used method for counting is a liquid scintillation counter (LSC) (Fig. 1). This device measures photons given off by photosensitive chemicals (fluors) in the liquid scintillation "cocktail" (a mixture of

Fig. 1 A representative, modern "bench-top" liquid scintillation counter, complete with printer. (Photo courtesy of Beckman Instruments, Fullerton, California.)

sample and solvents in a counting vial). While most counters have sophisticated computers and equipment to make final calculations of radioactivity easy, the principles of operation are both simple and elegant.

A liquid scintillation counter has paired photomultiplier (PM) tubes positioned 180° apart, facing a hole, or "well," which receives the sample vial. β^- particles emitted from the sample interact with the fluor molecules, creating flashes of light, which are detected by the PM tubes, which convert the light flashes into millivolt currents, which are then amplified, counted, and stored for display as "counts per minute." Because of the "4π" geometry (paired detectors monitoring virtually all the radiation in the sample vial) in the LSC, counting efficiency for ^{14}C in a high-quality counter can sometimes approach 98% of actual disintegrations per minute. An additional feature of having two detectors is a reduction in background. Unless both PM tubes "see" the flash, it does not register. Thus, spontaneous counts from only one do not register.

A few chapters, for example, those on carbon dating or tracing bomb-derived ^{14}C, address specific techniques for low-level counting. We present a brief account of generally used procedures for ^{14}C counting, and protocols for handling and safety. The essentials of planning and safe use of ^{14}C isotopes are presented in the protocol.

a. Sample Preparation   Various analytical instruments for processing samples of plant or animal material, soil, gases, etc. are used to obtain a "countable" sample, suitable for liquid scintillation counting. For example, one may have samples of gases (principally CO_2) evolved from ^{14}C-labeled leaf litter in a decomposition experiment. The CO_2 can be trapped by bubbling it in a CO_2-absorbing solvent (such as ethanolamine); then the solvent can be counted in a liquid scintillation counter.

Solid samples, for example, plant or animal tissues, can be dissolved directly in special solvents (which are rather costly), or they can be combusted in a carbon–hydrogen–nitrogen apparatus, such as a Dumas combustion apparatus, the volume of CO_2-carbon determined, the CO_2 absorbed in an absorbing solvent, and counted in an LSC, as noted above.

If one has obtained samples of CO_2 from soil respiration by trapping them in alkali, another method, called "suspension counting," is also very useful. Aliquots of the precipitated carbonate (CO_3^-) are pipetted into vials, and a gellike suspension of finely divided silica is added. The silica particles suspend the carbonate throughout the depth of the vial (typically 5–10 ml), and give very good counting efficiency. Gellike scintillation cocktails are also available, and are useful for suspension counting.

b. Quenching   Quenching is broadly defined as any decrease in efficiency of the energy transfer process in the scintillation solution. There are several types of quenching, principal among these being: (1) *chemical quenching,* where the substance being counted may interact with the excited molecules of the scintillation solution before they can emit the excitation as photons. Various nonfluorescing dissolved molecules, for example, polar compounds, such as alcohols, play this role, as they may absorb energy from the solvent molecules without emitting photons; (2) *dilution quenching,* where the fluor solution is markedly diluted by the volume of the sample to be counted; (3) *self-quenching* occurs if the primary solute molecule associates with un-excited solute molecules, thus decreasing energy in nonphoton production; and (4) *color quenching,* perhaps most common of all for biological samples, in which colored sample materials, for example, tissue pigments, absorb some of the fluorescence photons before they leave the counting vial.

Internal standardization and the automatic external standard are two of the most widely used techniques for quench correction.

1. Internal standardization   Although internal standardization is a reliable and accurate method of measurement of absolute counting efficiency, the technique is time consuming because every sample must be opened to insert a known amount of ^{14}C and counted a second time after the initial count. DPM values and counting efficiency can be calculated with the following equation:

$$E = [C_2 - C_1(100)]/D_{std}$$
$$D_{spl} = C_1(100)/E$$

where $E = \%$ counting efficiency of sample, $C_2 =$ counts per minute (CPM) of sample plus standard, $C_1 =$ CPM of sample alone, $D_{std} =$ DPM of standard, and $D_{spl} =$ DPM of sample.

2. External standardization   The Automatic External Standard (AES) is determined with a gamma emitter to produce Compton's electrons from the glass of the counting vial. The scintillation liquid reaches an excited state as a result of the electrons, and the level of energy transmitted by the liquid is compared to the expected level from the known gamma source. The ratio is the AES number and is always lower than 1. Standard curves should be determined for each different scintillation liquid. A plot of the counting efficiency as a function of the AES ratio should be linear in the range for calculating DPM values (Fig. 2). Linear regression equations for AES calibration curves should be standardized frequently to monitor the correct prepa-

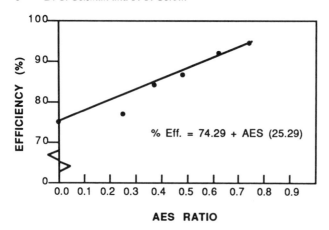

Fig. 2 A linear regression plot of the counting efficiency of a liquid scintillation spectrometer as a function of the automatic external standard (AES) ratio.

ration of scintillation cocktails and to assure accurate efficiency of the spectrometer.

II. SAFETY PRECAUTIONS

The main requirement for working with ^{14}C, as with all analytical studies, is a good quantitative technique. All pipetting should be done using a pipette bulb, or with disposable tips on a multiple pipettor. The usual protective gloves and lab gown or apron one would wear in a chemistry lab are *de rigueur.*

It is necessary (mandatory, in fact) to work in a clearly marked isotope working area, whose boundaries are delimited by radiation marker tape, with restricted access, away from the general traffic pattern in a laboratory area. Although ^{14}C is too weak an emitter to have any effects through glass or thin plastic, the main goal of *absolute minimal exposure* is to be observed at all times.

III. WASTE DISPOSAL

After conducting an experiment, there will be liquid and solid wastes to be disposed of. Any university, research laboratory, or other research facility will have a radiation safety officer, designated as safety supervisor of that particular laboratory. He/she will inform you of appropriate procedures. In general, liquid wastes are best kept in large, screw-capped plastic bottles (e.g., Nalgene bottles). Solid wastes (including paper towels, paper liner from lab bench working area) should be stored in lined cardboard boxes, clearly marked "radioactive waste." These can be disposed of later in a fashion authorized (supervised) by the radiation safety officer.

With increasing concerns about environmental hazards with buried wastes, it is preferable to use nontoxic fluors, to minimize aerial contamination in the laboratory.

ACKNOWLEDGMENTS

We thank J. Berg, W. Cheng, and D. A. Crossley, Jr. for reviewing the manuscript.

REFERENCES

Peng, C. T. (1981). "Sample Preparation in Liquid Scintillation Counting." Amersham Corp., Arlington Heights, IL.

Wang, C. H., Willis, D. L., and Loveland, W. D. (1975). "Radiotracer Methodology in the Biological, Environmental, and Physical Sciences." 480 pp. Prentice-Hall, Inc. Englewood Cliffs, NJ.

2

Photosynthesis/Translocation Studies in Terrestrial Ecosystems

F. R. Warembourg

Centre National de la Recherche Scientifique
CNRS Centre L. Emberger
F-34033 Montpellier Cedex, France

J. Kummerow

Department of Biology
College of Sciences
San Diego State University
San Diego, California 92182

I. INTRODUCTION

Biological systems are characterized by energy and matter fluxes. Energy available to green plants results from the conversion of light energy to chemical energy and is stored in reduced carbon units by the process of photosynthesis. This energy is then distributed within the organism by carbon translocation. Photosynthetic carbon becomes a part of multiple chemical combinations and a huge number of carbon compounds is synthesized in the processes of growth, reproduction, maintenance, and senescence of living structures. These activities cause the loss of energy and the release of CO_2 by respiration. This general model is valid not only for individual organisms but applies also to terrestrial ecosystems. It is not surprising that understanding the dynamic nature of organisms and ecosystems has greatly improved through the use of carbon isotopes, principally [14]C. The methods to investigate the many processes of accumulation, removal, and exchanges occurring in biological processes make use of the basic principles of isotope dilution and tracer kinetics. Developed initially for biochemical research of metabolic pathways, these methods have become an indispensable tool in plant physiology, agronomy, ecology, and soil science.

In this short chapter, the basic methods of ^{14}C use in plant science will be presented with three examples of applications in the field of plant physiology and ecology.

Since environmental factors play a major role in the rates of photosynthesis and translocation processes, a majority of the chapter is devoted to the description of methods and technologies involved to maintain normal growth conditions for the plants used for ^{14}C experiments. For complementary information on use of isotopes in photosynthesis and related plant functions, one may refer to the excellent reviews of Sestak et al. (1971) and Vose (1980). Translocation from plants to soil is well covered by Sauerbeck and Johnen (1971) and Martin and Kemp (1986).

II. MATERIALS REQUIRED

Experiments in photosynthesis and translocation studies of terrestrial vegetation using ^{14}C are of course specially designed for each problem to be addressed. However there are certain general features in common. Incorporation of ^{14}C is usually done via photosynthesis in the presence of $^{14}CO_2$. Measurements on plant parts, tissues, plant derived compounds, or soil follow a given translocation period. Therefore, basically all ^{14}C studies include the following items: (1) an exposure chamber, which causes a minimum disturbance to plants or plant parts and as little alteration of the environment as possible; (2) a system for delivering $^{14}CO_2$ in order to provide adequate ^{14}C at a rate suitable for normal plant carbon uptake and allowing later measurements of radiocarbon within the plant; (3) a sample procedure and sample preparation method designed to prevent any loss (e.g., by respiration; and (4) convenient sample analysis facilities for carbon and radioactivity measurements.

A.   Exposure Chamber

There is a great variety of labeling chambers in form, size, and type of material. This variety ranges from small leaf chambers for photosynthesis measurement (Tieszen et al., 1974) to large plastic tents designed for labeling herbaceous vegetation (Dahlman and Kucera, 1968). Exposure chambers can be simple enclosures or include sophisticated control equipment, dependent on the kind of investigation desired. Basically, an exposure chamber must be adapted to the type of plant or vegetation and to the location where it is to be used. This may be an open field, a greenhouse, or a laboratory. The chamber design must take into account the length of the exposure period, which may require adequate control of temperature, relative humidity, CO_2 concentration, efficient air mixing, and plant watering. For indoor units, artificial illumination is also necessary. The chamber must be gas proof in order to maintain an atmosphere enriched in $^{14}CO_2$, be transparent to light, and allow heat exchange in order to facilitate equilibrium between inside and outside conditions. Complex indoor phytotron chambers for ^{14}C exposure have also been designed to allow any desired environmental conditions (e.g., CO_2 level, temperature, illumination, and humidity) (Sauerbeck and Johnen, 1971; Andre *et al.*, 1974). Two kinds of exposure chambers will be considered for ecological studies.

1. Flexible Chambers

For short-term experiments on plants or plant parts growing in the field or in containers, the simplest exposure chamber consists of plain plastic bags (polyamide) provided with portholes for introducing or removing solutions or gas mixtures. As indicated in Fig. 1a and b, these bags are tightened around the stems or branches by means of a physiological mastic (Terostat, Teroson, Germany or Prestik, Blacking Corporation, Boston). The injection ports can be made with cut-off tops of plastic bottles (Fig. 1c), the covers of which are fitted with serum caps. This is an inexpensive and efficient way to provide an air-tight connection through the chamber wall. These plastic bags are especially convenient for ecological studies in the field, which require many simultaneous replicates. They are commercially available in various sizes to accommodate all kinds of plants. However, their flexibility may not always be convenient (e.g., woody or large-size plants). An important limitation is that no extra equipment can be adapted. Thus, these bags can be used only for pulse labeling experiments (a few hours maximum) with no environmental control.

2. Rigid Chambers

Most of the currently used chambers are made of rigid plastic material, generally perspex or plexiglass. Their construction requires some workshop facilities. An alternative is the use of a semirigid material such as Uvex (cellulose acetate butyrate) plastic (Goodfellow Metals LTD, Cambridge, England) (Warembourg *et al.*, 1982), which is easy to work with and presents good light transmission qualities. It can be formed under heat and sealed with acetone. The chamber illustrated in Fig. 2 is made out of a 1.5-mm thick sheet of this material. The dome shape is obtained by heating under pressure. Suitable for *in situ* labeling of natural vegetation, this kind of chamber may accommodate plants of a large range of sizes by adding a cylinder made of the same material at the base of the dome. Anchorage to the ground is achieved by tightening the chamber to a metal cylinder thoroughly inserted into the soil. A broad rubber band may serve as an airtight seal. Figure 3 illustrates a cylindrical chamber developed for the labeling of potted plants. The pots are sealed to a Uvex plastic base having a double-ringed side, which holds the edge of the chamber. When this rim is filled with water and a layer of oil on top, it ensures an airtight seal against the

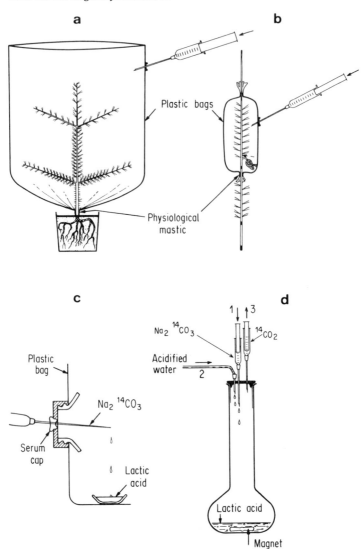

Fig. 1(a) and (b)   Flexible exposure chambers for labeling plants or plant parts with $^{14}CO_2$; (c) *in situ* generation of $^{14}CO_2$; (d) preparation of $^{14}CO_2$ air mixtures to be injected in exposure chambers. 1, 2, and 3, procedure steps.

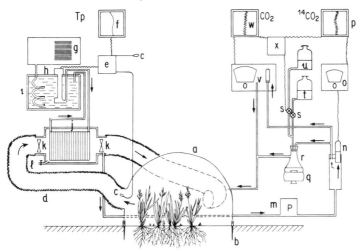

Fig. 2 Photosynthesis chamber, $^{14}CO_2$ generating, and temperature control equipment used *in situ*. a, Uvex chamber; b, stainless steel cylinder; c, temperature sensors; d, flexible tubing; e, temperature regulator; f, temperature recorder; g, refrigerator compressor; h, water pump; i, cooling bath; j, heat exchanger; k, induction fans; l, outlet for condensed water; m, air pump; n, Geiger–Müller probe; o, Geiger–Müller counter; p, radioactivity recorder; q, magnetic stirrer; r, reaction flask containing acid; s, solenoid valves; t, labeled carbonate; u, unlabeled carbonate; v, infrared CO_2 analyzer; w, CO_2 recorder; ac, regulator; Tp, temperature; P, pump.

outside atmosphere. The soil atmosphere can also be separated from the chamber atmosphere by using biological sealant (RTV silicone rubber, Dow Corning, Midland, Michigan) around the plant stems. Such a separation is a prerequisite for studying root and soil respiration. An alternative is to have the pots with air-tight covers located inside the chamber with a separate air circulation system (Johnen and Sauerbeck, 1977). Holes can be drilled through the walls of rigid chambers to allow for connections of various kinds of controlling devices and for circulation of the atmosphere in closed circuits (Fig. 2 and 3). Walls can be reinforced around the holes by gluing on with acetone a second layer of plastic material. These chambers may be used with or without environmental control equipment according to the length of exposure. They are convenient for use in the laboratory, greenhouse, or in open air.

B. Environmental Control inside the Chamber

For large-size chambers and long-term exposure to $^{14}CO_2$, it is desirable and frequently essential to control those environmental parameters that affect photosynthesis and plant growth. These environmental parameters include temperature, water vapor, atmospheric pressure in the chamber, light, soil

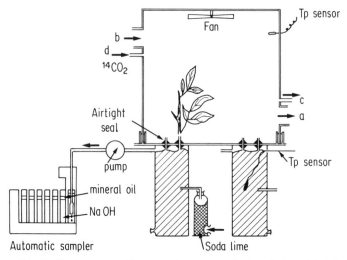

Fig. 3 Equipment used for measuring root respiration of plants labeled with $^{14}CO_2$: (a) and (b) connections with air mixing and temperature control equipment, (c) and (d) connection with $^{14}CO_2$ and CO_2 regulating equipment (see Fig. 2).

moisture, and the ambient CO_2 concentration. The latter will be discussed together with $^{14}CO_2$-generating equipment. These parameters can be kept constant or within narrow limits and can in many cases be maintained at predetermined values, although this is more a prerequisite for precise physiological work than for producing labeled material for soil biological studies (see Chapter 3, this volume). Detailed information concerning various means and devices for air conditioning of plant chambers may be found in books dealing with photosynthesis measurements (Sestak *et al.*, 1971). However, unlike photosynthesis studies in which air may be renewed periodically, $^{14}CO_2$ work has to be carried out in air-tight systems for the whole exposure period.

1. Control of Temperature

Techniques to regulate air temperature in the chamber depend upon its volume. For small chambers, the temperature may be regulated by mixing the air in the presence of heat exchangers located inside the chamber. This can be achieved with copper cooling coils mounted along the inside wall and connected to tap water or a source of liquid refrigerant. A temperature sensor may be used to control the circulation of this liquid (Warembourg and Paul, 1973).

For large chambers and especially for field work, commercial air-conditioning units can be used; however, it may be difficult to adapt them for

$^{14}CO_2$ work, which requires leak-proof equipment. As illustrated in Fig. 2, a system can be built at reasonable cost and used both for laboratory and field experiments. Temperature control within the chamber is achieved through a closed circuit system that includes a heat exchanger, flexible plastic tubing 10 cm in diameter, and the chamber. The heat exchanger consists of car radiators placed in an insulated airtight container (made of Uvex plastic) with two built-in fans. These radiators are cooled by refrigerant liquid (water + glycol) provided by a refrigeration unit. Such a unit consists of a compressor (3/4 hp refrigerator compressor), a cooling bath (large-size camping cooler of 50 l) with a built-in cooler coil, and a water pump that drives the refrigerant through insulated water tubing to the heat exchanger. In this system, air is permanently circulated by the fans from the chamber through the radiators and back. A temperature sensor located inside the chamber and connected to a thermostat helps control the flow of refrigerant from the cooling bath. The temperature is thus adjusted either by comparison to a reference or to the outside temperature indicated by a second sensor. The temperature can be maintained within 1°C of the outside temperature under any weather condition and for volumes up to 1 m³. With the exception of the refrigeration unit, the equipment is mobile and adaptable to various exposure chambers.

When labeling has to be maintained for several days, condensation is likely to occur inside the chamber at night when the cooling system is not operating. This problem can be resolved by inserting a heating element into the air circuit. In this case, air is continuously cooled and water is trapped at the heat exchanger maintained at the dew point temperature, while the temperature of the chamber is regulated by the operating heat source.

2. Control of Water Vapor

As already mentioned, water vapor, produced by soil evaporation and plant transpiration is condensed on the heat exchanger during the cooling period. To avoid extreme air dryness, the temperature of the heat exchanger should be maintained at around that of the dew point of the outside air unless extra equipment is included to regulate air humidity. A valve located in the lower part of the container holding the heat exchanger allows drainage of the condenser water at regular intervals. For pot culture experiments and long-term exposure this procedure may help to adjust soil moisture.

3. Control of Atmospheric Pressure

To prevent pressure changes that occur inside the chamber during cooling and heating, a large expansion bag (made of flexible plastic material or a car air bag) is included as a branch in the air circuit. Although pressure is not likely to affect the strength of a chamber made of semirigid material (Uvex), the worst effect of a pressure change that might occur while working *in situ* when soil and chamber atmosphere has not been separated is the following: CO_2 will be driven according to pressure changes from and to the soil, thus upsetting the CO_2 concentration and the specific activity of the air. A small air pump thus may become very handy in order to maintain a slight positive pressure in the system.

4. Control of Light

The necessity for supplying light arises only for indoor chambers. The requirement in artificial lighting is both qualitative and quantitative. The irradiance must be kept at a level and a spectral composition suitable for maintaining sturdy and healthy plants. The value on the surface of the earth at noon on a bright summer day at a latitude of $52°N$ is $400,000 mW/m^2$. In winter the figures are only about $1/10$ of this value. Experimental growth cabinets are normally designed to provide illuminance levels in the order of $25,000 - 35,000$ lux, equivalent to an irradiance of $75,000 - 100,000 mW/m^2$. Photosynthetically active radiation is in the range of 400 to 700 nm. This can be achieved with commercially available lamps, however, the amount of useful radiation per W/m^2 of installed power varies with the types. In general, artificial lighting is obtained using either a mixture of fluorescent and incandescent lamps or by high-pressure sodium or metal halide lamps, the efficiency of the latter being the highest. The number can be calculated according to the manufacturer's recommendations and the surface to be illuminated. For all types of lamps, a majority (60 – 80%) of the emitted power is released as heat, and cooling potential must be designed accordingly.

5. Control of Soil Moisture

For long-term labeling, which can in some experiments last several months, water availability to plants must be carefully controlled. This can be done accurately using tensiometers to monitor soil water. These consist of porous ceramic cups or tubes buried in the soil and connected to a mercury reservoir by small-diameter flexible tubing. The reservoir end of the tubing is mounted on a vertical stand provided with a graduated ruler, which allows measurement of water tension in centimeters of mercury. Providing calibration of the water tension with the water content of the soil has been done prior to the experiment, periodic adjustments can be made. This applies for single soil container or multiple pot experiments, each having its own tensiometer.

Water can also be automatically supplied according to the actual transpiration losses of each pot as determined by special balances, each of which governs an individual solenoid valve connected to a water tank (Johnen and

Sauerbeck, 1977). This of course requires that the pots be located inside the chamber.

C. System for Delivering $^{14}CO_2$ and Control of CO_2 Concentration

1. Source of CO_2

The $^{14}CO_2$ is generally prepared by the action of dilute acid (lactic or sulfuric acid) on $Ba^{14}CO_3$ or $Na^{14}CO_3$ aqueous solutions, the latter being generally preferred. These carbonates are commercially available in ampoules or flasks in sterile solutions of 40, 200, and 400 MBq with specific activities greater than 1 GBq/mmol. Appropriate dilution with unlabeled carbonate is then prepared by the investigator in order to meet specific requirements (e.g., volume of exposure chamber, CO_2 concentration, and length of exposure). $^{14}CO_2$ may also be purchased directly in gaseous form in "break seal" ampoules or in pressure tanks mixed with air.

2. System for Delivering $^{14}CO_2$ in Exposure Chambers

a. Pulse Labeling   In the absence of CO_2 partial pressure regulating equipment, $^{14}CO_2$ may be generated by injecting labeled carbonate with a hypodermic syringe through the injection port (Fig. 1c) into a vial with acid placed inside the chamber. In the field this procedure may be difficult to handle, therefore a system is used (Fig. 1a and b) in which the prepared $^{14}CO_2$ air mixture is directly injected into the chamber. This allows for quick injection into several chambers when replicates are needed. For this purpose, the required amount of labeled carbonate solution is injected into an Erlenmeyer flask of appropriate volume which contains lactic acid and is closed by a serum cap (Fig. 1d). Providing the flask has been evacuated prior to injection, stirring of the acid allows fast evolution of the CO_2 contained in the carbonate. A flask with acidified water is then connected to the Erlenmeyer flask by means of a hypodermic needle mounted on tygon tubing. This device is used to avoid an excessive vacuum and to replace the air volume taken up with a syringe for injection into the exposure chambers.

b. Continuous Labeling with Control of CO_2 Concentration   For large chambers and for long-term exposure, $^{14}CO_2$ must be generated continuously in order to maintain a known concentration of CO_2. This is achieved by a continuous flow system as shown in Fig. 2. Addition of labeled carbonate to acid generates $^{14}CO_2$, which is circulated by a diaphragm pump through a closed circuit independent of that used for temperature regulation. The system includes a reaction flask, a counting chamber, the pump, a continuous flow infrared gas analyzer, and connecting tubing (1-cm i.d.

tygon tubing). The reaction flask with the acid is connected to a solenoid valve and to a flask containing the labeled carbonate. The counting chamber consists of a glass tube with threaded ends, one of them mounted with a thin window Geiger–Müller probe connected to a rate meter. A constant flow of air is circulated by the pump between the exposure chamber and this circuit. The level of radioactivity and the CO_2 concentration are permanently monitored. A regulating device operates the solenoid valve on the carbonate flask when needed, responding to the signals from the CO_2 analyzer. The CO_2 concentration in the system can be maintained within 10 ppm of the chosen value.

This system is used when the only source of CO_2 is the added labeled carbonate, as is the case when the soil atmosphere is separated from the chamber atmosphere. In other situations and principally when working in the field, unlabeled CO_2 may evolve from the soil and dilute the ambient CO_2, thus lowering the specific activity. In some experiments, this specific activity must be maintained at a constant level. Therefore, a separate control of CO_2 concentration and radioactivity must be achieved. This is done by a double regulation system: one system operates by means of the Infrared Gaz Analyser (IRGA) control, which adds unlabeled carbonate, and the other system works through the Geiger–Müller counter-controls, which add labeled carbonate. Thus, CO_2 is labeled continuously *in situ* at a constant specific activity. When labeling for several days, a bypass system diverts the air flow to a CO_2 trap (ascarite or soda lime) included in a separate circuit to compensate for the increase of CO_2 during the night. During this time, radioactivity is not regulated to prevent an excess consumption of labeled carbonate. **Comment:** If available, a gas flow ionization chamber for monitoring and regulating the ^{14}C activity can easily replace the Geiger–Müller counting system (Sauerbeck and Johnen, 1971).

D. Sampling of Plant Material and Analysis

Following incorporation of ^{14}C and after a period of time specific to the investigation, plant material and/or plant-derived products are collected in order to analyze their content of [14]C. Generally, total carbon is measured also for specific activity determination.

1. Plant Material

The aboveground plant parts are harvested by clipping. They are carefully separated into categories (e.g., leaves, stems, and flowers) to meet the requirements of the experiment. Belowground parts are extracted and washed free of soil, using root-washing machines or thorough agitation of the roots in sieves of several mesh sizes in order to minimize root losses. In the field, root harvest is done by extracting soil cores. For pot experiments, the entire root system may be harvested. Fresh material may be either killed in liquid nitrogen and stored in a deep freezer for later analysis or dried at 70°C in an oven and stored in paper bags or envelopes. After being ground to powder, aliquots are burned by the dry combustion method and the resulting CO_2 is assayed for ^{14}C by scintillation counting (see Chapter 1, this volume). Direct counting of fresh plant material after digestion in acid mixtures has also been reported (Mahin and Lofberg, 1966). It applies only to young and nonfibrous material, and attention should be paid to colored compounds such as chlorophyll, which affect the counting efficiency. Comment: Although ^{14}C is one of the least hazardous radioisotopes, care must be taken when grinding plant material to prevent dust inhalation. A closed mill, located in a glove box or a well-aerated fume hood provided with air filters is required. A face mask must always be worn when manipulating the ground plant material.

2. Plant-Derived Compounds

Labeled $^{14}CO_2$ is the most frequently analyzed plant product because it is used to study metabolism and respiration as well as to control the ^{14}C content of the atmosphere to which the plants are exposed. It is generally absorbed in alkali (KOH or NaOH) or ethanolamine for further assay by scintillation counting. Individual gas samples are processed as follows: one ml ethanolamine or 0.2N NaOH solution is placed in a scintillation vial the top of which is closed by a serum cap. After producing a vacuum with a hypodermic syringe, 10 ml of the air to be analyzed is injected and left for a few hours. After addition of the scintillant, one proceeds with scintillation counting.

For continuous collection of CO_2 (e.g., from the soil atmosphere) an open circuit air flow technique is used as illustrated in Fig. 3. A flow of CO_2-free air (100 ml/min) is circulated by a pump through the root container. The CO_2 produced in this container is collected in NaOH solution. Aliquots are then assayed for ^{14}C and the remaining solution is analyzed for C contents by titration. The simplest CO_2 collector containing NaOH consists of a series of two gas wash bottles with fritted inlets (Sauerbeck and Johnen, 1971) or of two bubbling towers (a glass tube containing glass chips) tightly mounted on top of an Erlenmeyer flask (Sørensen, 1963) and connected to the aeration line. In both cases, the second unit acts as a safety for completeness of absorption. At intervals they are replaced by others with a fresh amount of NaOH. For the study of respiration dynamics an automatic sampler such as that illustrated in Fig. 3 may be used as a CO_2 collector, thus allowing numerous periodical measurements. To prevent contamination of the NaOH solution with ambient CO_2, a film of mineral oil is placed on the liquid surface of each collection tube, acting as a seal. Normality of the NaOH solution is set according to the respiratory activity of the studied material. For a normality higher than 0.2N, the volume of the aliquot used for scintillation counting must be determined according to the specifications given for the scintillant.

3. Chemical Extraction of Plant Material

For ecological and physiological studies, one may wish to know the product into which ^{14}C has been incorporated, be this structural or nonstructural, temporary or stable compounds. Biochemical and chemical analytical methods are used to quantify these compounds. These methods must lead to complete compound isolation in order to measure the ^{14}C content of each end product. This can be achieved by different methods often based on successive extractions, as indicated by a sample protocol (Fig. 4) adapted from Gordon *et al.* (1977).

One hundred milligrams of plant material dried and ground into fine powder is mixed in 20 ml ethanol (80% v/v) and boiled under reflux for 30 min. After filtration, the ethanol-soluble fraction is dried by rotary evaporation under vacuum and dissolved in 7.5 ml of water. The water soluble fraction is centrifuged (4000 rpm) and assayed for ^{14}C. It contains temporary metabolites: neutral compounds like sugars (e.g., glucose, fructose, sucrose) and charged compounds such as amino acids, organic acids, and sugar

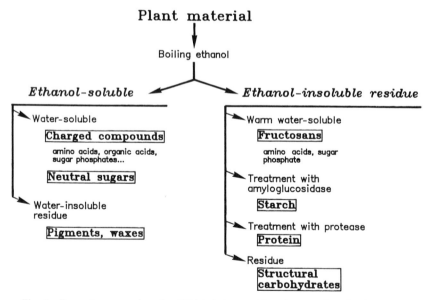

Fig. 4 Extraction procedure for ^{14}C-labeled material and categories of substances contained in various fractions. (Adapted from Gordon *et al.*, 1977.)

phosphates. The remaining residue, pigments, and waxes, is analyzed for ^{14}C after dry combustion and collection of $^{14}CO_2$.

The dried ethanol-insoluble fraction of the sample is extracted two times with 20 ml of water at 70°C. After centrifugation, the combined soluble extracts are assayed for ^{14}C. They contain fructosans, as well as some amino acids and sugar phosphates, insoluble in ethanol. The residue is incubated for 3 hr in 4 ml of boiling water in tubes provided with air condensers, then cooled, and starch is hydrolyzed with amyloglucosidase (*Rhizopus* mold, SIGMA Chemical Co., St. Louis, Missouri) at 55°C for 40 hr in a buffer solution with pH 4.5. The hydrolysate is assayed for ^{14}C content while the remaining starch-free residue is washed with water and treated with a protease (*Streptomyces griseus* Type XIV, SIGMA Chemical Co.) at 30°C for 24 hr in a buffer solution with pH 7.5. The centrifuged supernatant is assayed for ^{14}C; this fraction contains the amino-acid and peptide components of the proteins. The last residue is dried at 70°C and analyzed for C and ^{14}C after dry combustion. This material consists of structural carbohydrates such as cellulose and lignin. A ^{14}C balance is then calculated from the data of the various extracts and compared to the total ^{14}C content of the plant material. Recovery is higher than 95% if care is taken in recovering the material retained in filters. **Comment.** Drying causes chemical changes in the plant constituents and, according to the objectives, one may wish to use fresh material. After this has been killed in either boiling ethanol or liquid nitrogen, the same extraction protocol applies.

III. DESCRIPTION OF PROCEDURES

A. Pattern of Carbon Distribution at the Whole-Plant Level

1. Objectives and Principle

Carbon allocation patterns and carbon use in various plant parts have always been a challenge for plant ecologists and physiologists. This information is needed in order to estimate the cost of maintaining plant functions such as photosynthesis, mineral and water nutrition, and reproduction, not to mention the costs of parasitic and symbiotic interrelationships. The means of resource uptake, determined genetically and by environmental conditions, are associated with the costs of growth, differentiation, and function. These costs can be expressed in units of carbon invested in the respective plant compartments. Examples of these carbon allocation costs are: the ratios of aboveground to belowground production, vegetative to reproductive structures, ephemeral to perennial plant parts, and structural to storage carbon allocation. The use of ¹⁴C has proven very productive in this kind of investigation since ¹⁴C can be incorporated during a short period of time and traced in the plant components mentioned above. Comparisons can also be made between different plants, phenological stages, and environmental conditions. The methods use the basic principle of isotope dilution to establish the distribution of assimilates within the plant following a certain period of time after assimilation.

2. Materials

- Plants. Almost any kind of plant or plant parts are suitable for this type of investigation providing its size permits enclosure into a chamber. The plants, crop or wild plant, can be grown in pots or in the field.
- Simple $^{14}CO_2$ exposure equipment, either plastic bags or plastic cylinders suitable for simultaneous labeling, are used. For pot experiments, a large chamber capable of holding several individual plants can be an alternative.
- Previously prepared air mixture containing $^{14}CO_2$ and calculated to furnish ample isotope for the entire experiment, is the most practical means to label the plant or plant part with $^{14}CO_2$. However, labeled solutions to be injected into acid, contained in the exposure chamber, can also be used. The activity required for short pulse labeling is relatively high. For example, an exposure of 15 min in a 5-l chamber requires a level of activity of at least 1.5 MBq at a CO_2 concentration of 0.036–1%. It may be necessary to use an activity of 4–8 MBq if subsequent biochemical analysis is to be carried out. Knowing the required amount of $^{14}CO_2$, a solution can be prepared by calculating the needed quantities of Na_2CO_3, labeled and nonlabeled, using the information that the amount of CO_2 in one liter of atmospheric air equals 0.1928 mg, and the conversion factor for Na_2CO_3 to CO_2/C is 0.1463. The amount of unlabeled C in commercial solutions of labeled Na_2CO_3 is generally negligible compared to the amount needed to meet the CO_2 requirements in the chamber.
- Sampling equipment (scissors, root sampler) and facilities for sample preparation, extraction, and analysis as described previously are necessary.

3. Procedure

Carefully chosen healthy plants are enclosed in the bags, which are tightly sealed at the base of the plant or plant part. Five to ten replicates represent an appropriate number for statistical analysis. $^{14}CO_2$ is injected into each bag and left for 15 min. After this time, the remaining $^{14}CO_2$ can be partially removed by connecting the bag to a CO_2 collecting tower containing soda lime or ascarite. In the field, this is done by gentle hand squeezing of the bag. If possible, a diaphragm pump may be used to empty the chamber by circulating the inside air through the CO_2 trap. The plants are then left under normal growing conditions for a period of at least 4 days to allow for complete translocation of carbon from the exposed leaves to the other plant parts. After this period the plants are harvested, separated into categories (e.g., leaves, stems, roots), dried, weighed, ground, and analyzed for C and ^{14}C contents. Extraction of biochemical compounds may then be undertaken on aliquots.

4. Calculations

The data are calculated to express the amount of ^{14}C incorporated into plant parts or biochemical compounds using the equation:

$$^{14}\text{C in compartment A} = \frac{\text{Bqa}}{\text{Ca}} \times C_A = \text{Sp. Act.}_A \times C_A \qquad (1)$$

where

Bqa = disintegrations per second in aliquot a of A
Ca = carbon content in aliquot a
Sp. Act. = specific activity in Bq/mgC
C_A = amount of C in compartment A.

Summation of the activity recovered in each plant part makes it possible to estimate the proportions of distribution of assimilated C. According to the isotope dilution principle, these proportions apply to all the carbon assimilated during the same period.

Values of specific activities (Bq/mg C) are used to compare growth rates or activity rates of various sinks, being either different plant parts or biochemical components of a plant part.

An example will illustrate the type of calculation required before and after a $^{14}\text{CO}_2$ exposure experiment.

Let us suppose: (1) that carbon partitioning between organs of soybean plants, 70 days old, is to be estimated in the field, together with the proportion of carbon allocated into storage compounds as starch and (2) that the method chosen is a 15-min exposure of the plants to $^{14}\text{CO}_2$ using 20-l plastic bags. A scintillation counter is available for ^{14}C determinations.

At first, a rough calculation has to be made in order to determine the level of activity to be used for $^{14}\text{CO}_2$ exposure. Fifty to eighty days after sowing, soybean plants grown outside are in the linear growth phase with a net assimilation rate in the range of 300–500 mg of carbon per day. Assuming an average translocation of 20, 20, 17, and 3% of plant assimilates toward petioles, stems, roots, and nodules, respectively, 40% being retained by leaf blades during this period, one has to back-calculate for the smaller sink to be measured, that is, the nodules. Given a dry weight of 3 g and 5% of starch in nodules, adequate assay for radioactivity in aliquots of starch extracts should be at least 1.6 Bq for a safe detection in a scintillation counter. If extraction is done on 100 mg of nodules and assays on 1/10 aliquots, in considering the various dilution factors [extract aliquot (1/10), starch content (5%), nodule aliquot (100/3000 mg), and translocation to nodules (3% of assimilates)], the minimum radioactivity content of the whole plant should be

$$1.6 \times 10 \times 100/5 \times 3000/100 \times 100/3 = 0.3 \text{ MBq}$$

Given variations in biological material and losses during exposure, adequate amount of activity required inside the exposure bag should be at least 10 times this, or 3 MBq in a quantity of CO_2 suitable for insuring the minimum assimilation rate (7.6 mg C in 15 min, assuming a steady rate of 10 hr/day). This requires an initial CO_2 concentration of 0.1% or 10 mg C in the exposure bag.

Let us suppose that the following data are obtained from such an experiment. Respectively, for leaf blades, petioles, stems, flowers, roots, and nodules, the carbon content is: 3, 1, 2, 1, 2.5, and 0.5 g; the specific activity of each of them being 197.6, 265.2, 179.4, 93.6, 74.9, and 124.8 Bq/mg C. Total activity in plant parts and in the whole plant amounts to

$$592,800 + 265,220 + 358,800 + 93,600$$
$$+ 187,200 + 62,400 = 1,560,000 \text{ Bq}.$$

or

1.56 MBq

with a partitioning of 38, 17, 23, 6, 12, and 4% in the respective plant parts.

Given an initial CO_2 specific activity of 0.3 MBq/mg C, the 15-min net assimilation amounts to $1.56/0.3 = 5.2$ mg C. Wastes due to leaks and radioactivity left in the bag after removal plus amount lost by respiration can be estimated as $3 - 1.56$ MBq or about half the amount incorporated. The values of specific activity indicate that petioles and stems are, at that time, the most active plant parts in attracting assimilates.

If the radioactivity in aliquots of starch extracts is measured, the total amount of starch can be calculated for each organ and for the whole plant. For example: if a value of 100 Bq is found in starch extracts of nodules, starch in nodules represents

$$100 \times 10 \times 500/100 = 5000 \text{ Bq, or } 5000/62,400 = 8\%$$

of the carbon incorporated in nodules after exposure of the plants to $^{14}CO_2$ or 0.017 mg C.

5. Comments

The procedure of enclosing a plant or plant part in a plastic bag may cause physiological problems. First, the temperature will undoubtedly rise several degrees above ambient temperature and heat may damage the leaf tissue. Second, the CO_2 concentration will initially rise after injection of $^{14}CO_2$ and then gradually fall below ambient value. This will cause abnormal stomatal behavior and one cannot predict the assimilation rates. Extreme care should therefore be taken to avoid heat excess. It is preferable to run the experiment during relatively cool days with inside temperatures in the range of optimal temperature for the study plants. Carbon dioxide starvation may be avoided by reducing the length of the labeling period. However, the method is not meant to estimate absolute rates of CO_2 assimilation, but rather to allow incorporation of significant amounts of ^{14}C into the plants so that carbon transfers can be traced.

When available, a temperature- and CO_2- controlled chamber can be used for the same type of investigation. This would allow for extension of the labeling period, thus integrating possible hourly differences in translocation patterns. However, multiplication of such regulated chambers for simultaneous use in field replicates would be expensive.

B. Respiration Studies: Carbon Use in the Root System

1. Objectives and Principles

In biological systems, processes such as accumulation, removal, turnover, and exchanges occur. They are the results of growth, maintenance, senescence, or transfer, each process consuming energy. In terms of carbon, the loss of a certain number of units as CO_2 by respiration is involved. The origin of the respired carbon in complex systems is therefore variable and investigation of any of these processes through respiratory activities is difficult because quite often they take place simultaneously. In a soil, the situation is even more complex since CO_2 production is the result of plant root activities as well as many other organisms' activities. Root respiration is due to root growth mainly supported by current photosynthesis and root maintenance partially in charge of the renewal of labile structures. Other organisms' respiration consists of symbiotic and/or associative microbial activities, which rely on plant carbon either directly or through root-deposited carbon. It also includes activities associated with the microbial breakdown of soil organic matter as well as faunal activity. The sources of carbon in soil respiration are therefore numerous but most important, they are of different age, and distinctions can be made once ^{14}C has been introduced into the system. Plant-derived and soil-derived carbon losses can be separated if the plants have been exposed uniformly to $^{14}CO_2$ from the seedling stage. Concepts and technical approaches concerning long-term labeling of plants can be found in Sauerbeck and Johnen (1971) and Johnen and Sauerbeck (1977).

Further detailed information can be gathered when ^{14}C is introduced into the plant system by photosynthesis during a short period of time. One may obtain a time sequence of the processes involved by analyzing ^{14}C appearance in soil CO_2. The objective of such a kinetic analysis is to identify processes, determine transfer rates of carbon, and estimate quantities. This analytical method has been proven extremely useful for the study of root respiration (Ryle et al., 1976), microbial respiration in the rhizosphere (mainly legumes and grasses) (Warembourg and Billes, 1979; Warembourg, 1983), but also for the investigation of soil organic matter (see Chapter 5, this volume). The method could well be designed for other plant parts. The basic protocol presented here uses the example of short-term exposure to $^{14}CO_2$ for studying respiration associated with root activities.

2. Materials

- A two-compartment system similar to that illustrated in Fig. 3 is used for feeding the plants with $^{14}CO_2$, separating aboveground and belowground plant parts, and collecting soil CO_2. It must accommodate several plant containers to allow replicate measurements or multitreatment studies. This system should include a continuous $^{14}CO_2$-generating device with $^{14}CO_2$ regulation and temperature control for long-term exposures. Control of CO_2-specific activity inside the chamber may be useful.
- The equipment used for collecting the CO_2 from the soil consists of an aeration train as illustrated in Fig. 3. A peristaltic pump with individual pump heads for each of the plant containers helps circulate the air. Instead of an automatic sampler to collect the respired CO_2 as illustrated in the figure, the collector can consist of bubbling towers or gas wash bottles as indicated above. A titration apparatus (pH titrator or manual) is needed to measure the collected CO_2.
- The source of $^{14}CO_2$ consists of a solution of labeled sodium carbonate, 10–40 MBq/g C according to the length of the exposure period (several hours to less than one hour).
- Other reagents consist of a 0.2N or less concentrated solution of NaOH for CO_2 collection, depending on the respiratory activity, 0.1N HCl for titration, $BaCl_2$, and scintillant for aqueous solutions (e.g., Scintillator 299, Instagel, Packard Instruments, Downers Grove, Illinois).

Analytical facilities for C and ^{14}C measurements in plant material are also required.

3. Procedure

a. Exposure to $^{14}CO_2$. The plants (either different or with different treatments, according to the type of investigation) are grown in pots or plastic containers provided with properly placed openings (Fig. 3). Before exposure to ^{14}C, the base of the chamber, pierced with holes for inserting the aboveground part of the plants, is tightly attached to the root containers using biological rubber sealant or physiological mastic (Terostat, or Prestik). A joint of this sealant is then placed around the plant stems. When a catalyst (RTV silicone rubber) is used, a support of glass wool must be first inserted into the hole in order to hold the liquid material. Now one proceeds with exposure to $^{14}CO_2$ from less than one hour to one day. Afterward, the $^{14}CO_2$ is replaced by nonlabeled CO_2 and the chamber is removed.

b. Respiration measurements. During and after exposure to ^{14}C, CO_2 is collected from the root containers and absorbed in NaOH solutions that are replaced periodically (every 1–4 hr, according to the frequency of information needed). An aliquot of 1 ml (or less, according to the normality of the sodium hydroxide and the specification of the scintillant used) is assayed for ^{14}C, the remaining being titrated for C content (the reacted NaOH is precipitated with $BaCl_2$ and the unreacted NaOH titrated with HCl). Measurements are continued for several days up to one week.

c. Sampling of plant material. After this period, the plants are harvested, separated into categories (e.g., aboveground, roots, nodules), and analyzed for C and ^{14}C.

4. Calculations

The amounts of $^{14}CO_2$, CO_2, and specific activity are calculated from ^{14}C and C measurements of the collected CO_2. When plotted against time the various data indicate the dynamics of labeled and nonlabeled CO_2 evolution. Qualitative and quantitative comparisons can then be made between different plants or treatments. Mathematical analysis can also be helpful in order to investigate the time course of different processes. Since the fractional rate of change of many biochemical processes is constant, the production of CO_2 by one of them after a short exposure of the plant to $^{14}CO_2$ will be exponential. The use the semilogarithmic paper to plot the data may therefore aid in separating successive CO_2-producing processes such as biosynthesis and maintenance of root structures. Cumulative amounts of C are compared to the amount recovered from the roots. This expresses the fraction of carbon loss in respiration.

5. Comments

Application of this analytical method may be extended to more complex situations in which the root system is supporting microbial populations, symbiotic or not, which impose an important stress on the roots such as nitrogen fixers (Warembourg, 1983) or mycorrhizae (Pang and Paul, 1980; see also Chapter 3 by Harris and Paul, this volume). Comparison of inoculated and noninoculated plants helps to assess the activity associated with these microorganisms. Mathematical analysis of the time course of $^{14}CO_2$ evolution and identification of the various respiratory processes is facilitated when the introduction time of the isotope is reduced (Warembourg and Billes, 1979). A great deal of fundamental work still must be done in order to establish a standard methodology for separating satisfactorily root and microbial activities in the rhizosphere and so far, quantitative comparisons of CO_2 effluxes are the most common expression of the microbial influence on the entire respiratory process.

One should be aware that the success of detailed analysis of root and soil $^{14}CO_2$ evolution after labeling the plant with $^{14}CO_2$ depends on the labeling conditions. Even for short-term labeling, CO_2 and specific activities must be kept constant so that variations in the $^{14}CO_2$ effluxes can be attributed only to internal processes. This requires reliable regulation equipment.

In the field of soil science, the use of ^{14}C has been subjected to further developments that go far beyond the estimation of root respiration. This is due to the great interest in the extent of plant root effects on the carbon and energy status of the soil. Often called the "rhizosphere effect," it is known to be the result of carbon release other than CO_2 by plant roots. It ranges from passive leakage to active secretion of carbon compounds in the vicinity of living roots. Additional contributions come from root cells and tissues deposited in the soil after senescence. Therefore, many additional ecosystem aspects have been investigated with ^{14}C in order to estimate the role of such processes on both the carbon balance of the plant and that of the soil. Complementary to the estimation of root and microbial respiration is, for example, the release of organic carbon, the use of this root-derived carbon by soil organisms, and the effect of this on the simultaneous breakdown of soil organic matter. Methods and techniques have been designed for sampling and investigating the rhizosphere soil (Helal and Sauerbeck, 1983), for distinguishing between root and microbial respiration using antimicrobial agents (Martin and Kemp, 1986; Helal and Sauerbeck, 1987), and for estimating the turnover of roots (Johnen and Sauerbeck, 1977).

C. Carbon Gain, Distribution, and Fate in Terrestrial Ecosystems: Field Studies

1. Objectives and Principles

If the rate of photosynthesis is the first important factor in the production process, carbon partitioning, efficiency of assimilate conversion, and residence time within plant parts are of equal if not greater importance. To understand ecosystem functioning, solid knowledge of the real amount of carbon that circulates within the system is needed. It requires an estimate of carbon gain, distribution, and losses during a given period of time in order to calculate the net production and therefore the real amount of material allocated at each trophic level. ^{14}C has been used as a means to estimate *in situ* assimilate distribution, secondary transfers, production, and senescence of various plant parts during a growing season because it can be introduced at a certain time and traced into the system. The methodology is again based on the principle of isotope dilution, assuming that in any compartment, a small amount of radioisotope is as evenly distributed as the nonlabeled isotope.

2. Materials

- Plant. The *in situ* study of carbon dynamics can be undertaken on natural herbaceous vegetation, crop plants, or on any soil suitable for chamber positioning and soil coring. Plots are chosen to accommodate several labeling experiments and successive harvests (5 × 5 m, for example). Fencing must be provided to prevent consumption of labeled material by animals. The proximity of a power outlet is also highly desirable.
- The exposure chamber and the regulating equipment are similar to those described previously (Fig. 2). The regulating equipment must be kept under cover (trailer, truck, or garden shelter) with an adequate power supply.
- Sampling equipment is required. Soil coring can be done with manual or hydraulic augers. The required sample processing equipment and analytical facilities are those described previously.
- The $^{14}CO_2$ source consists of a prepared solution of labeled sodium carbonate with a specific activity of approximately 10 MBq/g C. Other reagents include diluted acid (lactic or sulfuric diluted in two volumes of

water) for $^{14}CO_2$ production and those necessary for sample analysis (see above).

3. Procedure

a. Exposure to $^{14}CO_2$. The metal cylinder used to attach the exposure chamber is positioned on the ground and driven at least 10 cm deep into the soil to prevent leakage of the atmosphere. It is preferable to position this ring several days in advance before labeling in order to minimize disturbance of the vegetation. Some water may be used outside the cylinder to ease its insertion into the ground. For dense vegetation, adequate access (e.g., a boardwalk) must be provided in order to protect the surrounding plant cover.

Exposure to $^{14}CO_2$ is performed on a sunny day. The chamber is attached to the metal cylinder and the various regulating devices set up. Temperature regulation is started as soon as the chamber has been installed, followed by the CO_2 and $^{14}CO_2$ feeding and regulation. Exposure is maintained during one photoperiod (one full day) after which $^{14}CO_2$ is removed by circulating the chamber air in a CO_2-absorbing reagent (soda lime or ascarite). The equipment is then dismantled and the plants are left under natural growth conditions. Successive exposures at different areas are performed throughout the growing season. Plant phenology may serve as a guide for the choice of the labeling periods.

b. Sampling of plant material. One week after exposure, one proceeds to the first harvest of plant material. Sampling is done on one quarter of the labeled area. The remaining three quarters remain intact for later sampling. The sampling dates during the season are chosen to match the schedule of the labeling experiments. Thus, sampling 2 of labeled area 1 is performed at the same time as sampling 1 of labeled area 2, and so on in the case of 4 labeling experiments. Aerial plant material is clipped and separated into categories (for grasses: leaves, sheaths, stems, reproductive structures, and dead material). Care should be taken to separate senescent material, mainly in the leaf category, since this will help to evaluate the rate of senescence and aboveground growth dynamics as shown below. Belowground plant parts are extracted from soil cores, carefully freed from soil by washing, and in the case of grasses, separated into sheath and stem bases, rhizomes, litter, and roots at different depths (cores are cut into 10-cm segments). A minimum of 5 cores must be taken from each sampled area if statistical analysis is to be meaningful.

c. Sample analysis. Samples of each category are then dried, weighed, and ground. Analyses of the ^{14}C and C contents are done on aliquots using the standard methods described above.

4. Calculations

a. Assimilates distribution. The amounts of labeled C recovered from each plant part and expressed per unit area (m²) are used to calculate the total amount of labeled carbon stored in the vegetation at each sampling date. One week after ¹⁴C incorporation, all the respiration associated with the synthesis of new structures has occurred (McCree, 1974; Ryle *et al.*, 1976). Therefore, primary translocation and synthesis has taken place, and the amount of ¹⁴C recovered in each plant part at that time represents the net production of labeled compounds consecutive to the day of exposure. According to the principle of isotope dilution, the relative proportions thus obtained represent the distribution pattern of total net production prevailing at the period of labeling. The same applies to the data obtained for the material harvested one week after each successive labeling period. Calculations are those of procedure A, 4. (see page 25).

b. Fate of carbon, secondary transfers, and respiration. The amount of ¹⁴C recovered from each plant part after increasing time periods following the first sampling are compared to the amounts measured previously. Since senescent and otherwise broken-off plant parts are collected, the decrease in total plant ¹⁴C content is attributed to losses by respiration. The rate of decrease can therefore be calculated for each period between harvests. Thus, as a first approximation, a coefficient of loss due to respiration can be attributed to each plant part according to its initial content of ¹⁴C. Within these categories, changes are due to respiration but also to remobilization to other organs. Translocation coefficients can then be calculated as the difference between the measured change in ¹⁴C content and the loss associated with respiration. These coefficients are negative for compartments that export C, positive for the others. Enrichment of dead leaves (yellow or brown) in ¹⁴C indicates the rate of senescence for green leaves.

c. Production of different plant parts. The model for calculating production is based on two equations derived earlier by Wareing and Patrick (1975):

$$Pt = PA + PB + \ldots PN, \tag{2}$$

that is, total production (Pt) of a system is the sum of the production of each compartment (A, B, . . . N), and

$$A = PbA - (R + T + RT), \tag{3}$$

that is, balance of C around a single compartment (A) is equal to input by photosynthesis or import of assimilates (gross production, PbA) minus output by primary respiration (R), translocation to other plant parts and senescence (T), and respiration associated with these transfers (RT).

From Eq. (2),

$$Pt = aPt + bPt + \ldots nPt \tag{4}$$

with $a + b + \ldots n = 1$ being the coefficients of distribution of the production in each compartment, and

$$Pt = PA/a. \tag{5}$$

From Eq. (3),

$$PbA - R = PA, \tag{6}$$

the net production of compartment A. Rearrangement between Eq. (2) and (3) gives

$$PA = A + (T + RT); \tag{7}$$

therefore, knowing (1) the biomass changes with time and rate of losses of a single compartment, and (2) the distribution pattern of net production within each compartment, it is possible to estimate the production of each compartment and therefore total production.

Changes of green leaf biomass over time are obtained by periodic sampling. Estimates of distribution, secondary translocation, respiration of assimilates, and senescence coefficients are made by following the dynamics of ^{14}C concentrations over the same periods after incorporation.

With all the parameters being measured or calculated, the production of green leaves can be estimated day by day using a numerical series of the type:

$$\text{from } t = 0 \text{ to } t = n$$

$$PA = lA(t - 1)t/2 + PAt_1(t) \tag{8}$$

where $l = (T + RT)$, the daily rate of loss, and $PAt_1 = $ the production on day 1.

d. Turnover of root C. In other plant parts, the change in the ^{14}C amounts measured in the root compartment over a period of time is used to calculate root carbon disappearance or turnover after correction is made for possible remobilization to other plant parts.

5. Comments

One of the main problems of data interpretation is the heterogeneity of biomass values. In fact, this problem is well known to ecologists: biomass estimates, mainly for underground plant parts, require tedious sampling and more replicates than those obtained from the labeled areas. The same applies for successive samplings on one area labeled with $^{14}CO_2$. The area under the chamber may be too small to provide enough replicates and therefore meaningful statistical analysis. Furthermore, in vegetation types with horizontally extending rhizomes, extraction of soil cores may affect the adjacent vegetation. The same is valid for the removal of one part of the vegetation cover. To avoid these undesired side effects and increase the sampling area, several sites corresponding to the number of projected samplings may be labeled successively at 1-day intervals.

IV. GENERAL COMMENTS

A. Pulse/Continuous Labeling Procedures

Controversies often arise concerning nonhomogeneous versus homogeneous labeling of plants when planning and interpreting ^{14}C experiments. In fact, it all depends on the question asked and on the system under investigation. To avoid confusion, one has to refer to the basic principles of tracer use in dynamic studies.

1. By definition, the tracer is meant to trace the destiny of the element or compound into which it has been mixed or incorporated and with which it moves through the various components of a given system. It is therefore desirable that the tracer becomes incorporated at one location of this system during a short period of time so that the whole system does not become homogeneously labeled.
2. If a tracer is used to estimate the transfer of a compound from one compartment to another, the tracer must be evenly distributed in the compound it is supposed to trace. This applies in all cases and therefore in all studies involving ^{14}C as a tracer.

Thus, a pulse or short-term labeling of plants or vegetation is meant to trace the fate of carbon assimilated during a given period of time. During this time, the carbon that is incorporated must be homogeneously labeled with ^{14}C and later, when transferred from one compartment to another (e.g., plant parts or biochemical compounds), it must be evenly distributed. This does not mean that all the components of a compartment nor all compartments are evenly labeled.

If a distinction must be made between soil-derived and plant-derived carbon for estimating the overall root allocation to the rhizosphere, it requires that all the carbon incorporated in the plants be homogeneously labeled with ^{14}C. This means uniform long-term labeling of plants during their whole growth period. The same applies in studies of the fate of plant material, such as in decomposition studies, where labeled material is incorporated into the soil at a given time and is then transferred from one compartment to the other through decomposition processes. It therefore has to be homogeneously labeled with ^{14}C in order to give meaningful estimates and then evenly distributed in each compartment where it must be measured. At the time scale of decomposition processes, the introduction of labeled plant material is similar to a pulse labeling of the soil system.

B. Level of Radioactivity

One may be concerned about the occurrence of plant injury following exposure to ^{14}C radiation. Sauerbeck and Fuhr (1963) have determined that plants of several species grown for long periods of time (6 months) exposed to activities of up to 1 MBq/g C grew very well but the seeds showed declining viability. At 10 MBq/g C growth was affected. These results indicate that short-term labeling at the levels of activities recommended in the preceding protocols will have no effect on plant behavior.

REFERENCES

Andre, M., Nervi, J. C., Lespinat, P., and Massimino, O. (1974). Units for automatic culture in artificial atmosphere. *Acta Horticulturae* 39, 59–72.

Dahlman, R. C., and Kucera, C. L. (1968). Tagging native grassland vegetation with carbon-14. *Ecology* 49, 1199–1203.

Gordon, A. J., Ryle G. J. A., and Powell, C. E. (1977). The strategy of carbon utilization in uniculm barley. 1. The chemical fate of photosynthetically assimilated ^{14}C. *J. Exp. Bot.* 28, 1258–1369.

Helal, H. M., and Sauerbeck, D. R. (1983). Method to study turnover processes in soil layers of different proximity to roots. *Soil Biol. Biochem.* 15, 223–246.

Helal, H. M., and Sauerbeck, D. R. (1987). Methods of sampling and investigation of rhizosphere soil. In *Proc. 20th Colloqu. Int. Potash Institute, Bern*, pp. 235–246.

Johnen, B. G., and Sauerbeck, D. R. (1977). A tracer technique for measuring growth, mass and microbial breakdown of plant roots during vegetation. In "Soil Organisms as Components of Ecosystem." *Ecol. Bull. (Stockholm)* 25, 366–373.

McCree, K. J. (1974). Equations for the rate of dark respiration of white clover and grain sorghum, as function of dry weight, photosynthetic rate and temperature. *Crop Sci.* 14, 509–514.

Mahin, D. T., and Lofberg, R. T. (1966). A simplified method of sample preparation of tritium, ^{14}C, ^{35}S in blood or tissue by liquid scintillation counting. *Anal. Biochem.* 16, 500–509.

Martin, J. K., and Kemp, J. R. (1986). The measurement of C transfers within the rhizosphere of wheat grown in field plots. *Soil Biol. Biochem.* 18, 103–107.

Pang, P. C., and Paul, E. A. (1980). Effects of vesicular-arbuscular mycorrhize on ^{14}C and ^{15}N distribution in nodulated fababeans. *Can. J. Soil Sci.* 60, 241–250.

Ryle, G. J. A., Cobby, J. M., and Powell, C. E. (1976). Synthetic and maintenance respiratory losses of $^{14}CO_2$ in uniculm barley and maize. *Ann. Bot.* 40, 571–586.

Sauerbeck, D. R., and Fuhr, F. (1963). Experiences on labelling whole plants with carbon-14. In: FAO/IAEA, The use of isotopes in soil organic matter studies. *Brunswick Volkenrode*, pp. 391–398.

Sauerbeck, D. R., and Johnen, B. G. (1971). Ein vollstandiges Klein-Phytotron zur Herstellung ^{14}C markierter Pflanzen. In *Proc. 8th Int. Symp. Nuclear Energy and Agriculture, Pisa*, pp. 217–228.

Sestak, Z., Catsky, J., and Jarvis, P. G. (1971). "Plant photosynthetic production, Manual of methods." 818 pp. W. Junk N. V. Publishers, The Hague.

Sørensen, H. (1963). Studies on the decomposition of ^{14}C labelled barley straw in soil. *Soil Sci.* **95**, 45–51.

Tieszen, L. L., Johnson, D. A., and Caldwell, M. M. (1974). A portable system for the measurement of photosynthesis using ^{14}Carbon dioxide. *Photosynthetica* **8**, 151–160.

Vose, P. B. (1980). "Introduction to nuclear techniques in agronomy and plant biology." 391 pp. Pergamon Press, Oxford, England.

Wareing, P. F., and Patrick, J. (1975). Source-sink relations and the partition of assimilates in the plant. pp. 481–499. "Photosynthesis and productivity in different environments (J. P. Cooper, ed.) Cambridge Univ. Press, Cambridge, England.

Warembourg, F. R. (1983). Estimating the true cost of dinitrogen fixation by nodulated plants in undistributed conditions. *Can. J. Microb.* **29**, 930–937.

Warembourg, F. R., and Billes, G. (1979). Estimating carbon transfers in the plant rhizosphere. In "The soil-root interface," pp. 183–196. (J. L. Harley and R. S. Russell, eds.). Academic Press, London.

Warembourg, F. R., Montange, D., and Bardin, R. (1982). The simultaneous use of $^{14}CO_2$ and $^{15}N_2$ labelling techniques to study the carbon and nitrogen economy of legumes grown under natural conditions. *Physiol. Plant.* **56**, 46–55.

Warembourg, F. R., and Paul, E. A. (1973). The use of $^{14}CO_2$ canopy techniques for measuring carbon transfer through the plant-soil system. *Plant and Soil* **38**, 331–345.

3

Techniques for Examining the Carbon Relationships of Plant – Microbial Symbioses

D. Harris
E. A. Paul

Department of Crop and Soil Sciences
Michigan State University
East Lansing, Michigan 48824

I. INTRODUCTION

A. Symbioses in Which ^{14}C Is a Useful Tracer of C Flow

Symbioses between higher plant root systems and microorganisms frequently involve the exchange of energy in the form of reduced carbon compounds from photosynthesis for inorganic nutrients. Biological N fixation, particularly by legume – *Rhizobium* symbiosis, has received much attention because of its importance to agriculture and to global N cycling and its potentially high cost in terms of host plant photosynthate. Less attention has been paid to the C costs of mycorrhizal symbioses despite their extremely widespread occurrence in nature. Vesicular-arbuscular (VA) mycorrhizal fungi appear to be obligate biotrophs, deriving all their C from the host plant. While it is clear that ectomycorrhizal fungi also derive C from the host, the extent of saprophytic activity and thus the proportion of fungal C derived from the host plant is uncertain. Other looser associations of rhizosphere and rhizoplane organisms such as *Azospirillum* and various associative N-fixing organisms have no specialized structures for the exchange of materials with the host plant, but may be biochemically specialized to regulate plant growth and activity.

B.   Carbon Costs to Plant

1. Theoretical Estimates

The broad limits of C costs to plants hosting microbial symbionts can be estimated by consideration of the biomass and expected molar growth yields of the various components of the symbiosis. Consider, for example, a hypothetical soybean plant that grows from a 0.1 g seed to achieve maximum biomass after 12 weeks. In the absence of symbionts a total plant biomass of 30 g (12 g C) is formed from the photosynthetic fixation of 20 g C. The molar growth yield of this plant is 0.6 and its average specific growth rate falls between the limits 0.14 and 0.07/day, representing linear and exponential growth. We now add rhizobial and VA mycorrhizal symbionts to this plant and stipulate that 50% of its total N content of 1 g is derived from N fixation. The biomass of root nodules and mycorrhizal fungus at 12 weeks are both 0.5 g (0.2 g C). If the molar growth yield of nodules is assumed to be 0.5 and the ratio of C used to N fixed by nitrogenase is 3 : 1 (Atkins *et al.*, 1978), the C diverted to root nodules will be 1.9 g, of which 0.4 g will be used in symbiont growth and 1.5 g by nitrogenase. The molar growth yield of VA mycorrhizal fungi appears to be low (Harris and Paul, 1987) (here we will assume a value of 0.2); thus, mycorrhizal fungus growth would require 1 g C. If plant biomass C is still to achieve 12 g, total photosynthesis must be increased to 22.9 g C, reducing plant molar growth yield to 0.52, a decrease of 13%. In the absence of increased photosynthesis, the decrease in molar growth yield would reduce the average specific growth rate by the same 13% proportion. From this we can place approximate limits on the potential effects of the C costs of these symbioses in reducing plant biomass, from 13% if growth were linear, to 45% if growth were exponential.

In real symbiotic systems many other factors must be considered. Important among these are the actual limitations to plant and symbiont growth in a particular environment and the nature of growth curves of the component partners of the symbiosis. When plant photosynthesis and C export are limited by nutrient availability or by feedback inhibition, there is potential for increased C fixation in the host which may "compensate" for the C cost of the symbiosis if an effect of the symbiosis is to relieve the limitation. This may occur through an increase in the supply of the limiting factor, for example P or N, or simply by the microbial utilization of excess carbohydrate. Symbioses may also alter the architecture of the host plant in ways that increase photosynthetic capacity through increases in the ratio of leaf area to total biomass (Harris *et al.*, 1985).

The allocation of C to microbial symbionts is not uniform throughout the growth and activity cycles of the component organisms. The mutualistic effects of C and nutrient exchange depend therefore on the synchrony of growth cycles as well as edaphic and environmental factors. Tracer techniques can be used to measure these costs.

2. Measurements

The theory of photosynthetic pulse-chase labeling as a method of examining the C economy of systems and the general methodology of exposing plants to pulses of $^{14}CO_2$ have been described by Warembourg and Kummerow, in Chapter 2, this volume. We will deal with methodologies important to the examination of symbioses and point out features of experiment and equipment design that we have found to be important in some of the symbiotic relationships examined over the last 20 years.

II. MATERIALS AND PROCEDURES

A. Chamber Design

Much of the carbon flux through the symbiotic root system will be released as respired CO_2; it is therefore important to design the exposure system to separate and contain both the aboveground and belowground atmospheres. Separation of the belowground atmosphere can be achieved by growing the plants in cylinders that can be closed at the top and bottom. This requires a gas-tight seal around the plant stem. Terostat or some brands of modeling clay (Permoplast, American Art Clay Co., Indianapolis, Indiana) work very well as stem sealants for woody or herbaceous plants with distinct primary stems; grasses pose severe problems because the seal must usually be made above the first node and gas leaks between the inner leaves are frequent. Polyvinyl chloride (PVC) drain pipe is inexpensive, locally available in many diameters, easily machined, and can be glued readily to itself. It is thus a good material for the construction of belowground containers. Cylinders may be split longitudinally and clamped or bolted together to facilitate dissection of the soil – plant – symbiont system at the destructive harvest of the experiment. A typical apparatus for exposing plants and measuring or collecting respired CO_2 is shown in Fig. 1.

Cylinder size must be a compromise between sufficient rooting volume and the difficulty of harvesting root and other biological components from large volumes of soil or other media. In cylinders of any reasonable size roots will inevitably be concentrated at the cylinder walls. *In situ* experiments have been conducted in undisturbed soil blocks (1 m³) made with a narrow ditching machine ("Ditch Witch"). Plywood dividers were inserted into a square array of slots in the field soil and the vertical faces of the monoliths wrapped in 0.18 mm vinyl sheet before back-filling the slots.

Large volumes of soil can reduce the sensitivity and dynamic accuracy of measurements of the respiration of the plant–symbiont system by (at least) two mechanisms. First, by buffering the exchange of respired $^{14}CO_2$ with the collection system and second, by diluting the plant–symbiont-derived CO_2 with CO_2 from mineralization of organic matter in the soil. Most experiments on symbiotic systems will involve comparisons between plants of

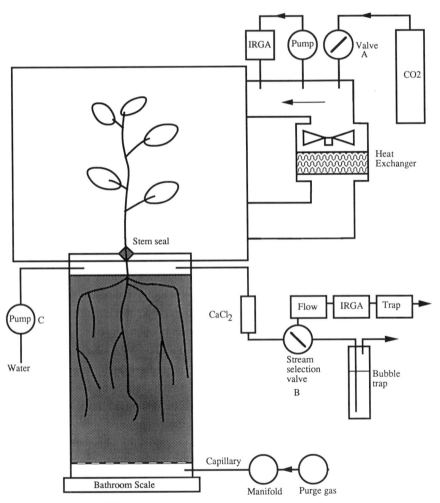

Fig. 1 Diagram of $^{14}CO_2$ labeling and respiration chamber. Analog signals for aboveground and belowground CO_2 concentrations, belowground purge flow rate, canopy air temperature, and cylinder weight are digitized and read by a microcomputer. The program controls canopy CO_2 concentration by admitting $^{14}CO_2$ through valve A and valve B sequentially selects each of several purge flows. Water is added to cylinders by pump C.

different symbiotic states so that several plants will be needed for each experiment. It is expensive to duplicate the control apparatus for the aboveground atmosphere of each plant. We use an aboveground chamber common to all plants. This requires only a single set of environmental control devices and ensures that all plants are exposed to the same aboveground atmosphere and environment. Disadvantages are that CO_2 consumption by photosynthesis and production by shoot dark respiration cannot be attributed to individual plants.

We fabricate aboveground canopies from a variety of transparent plastic films depending on the required size and application. Small, self-supporting canopies for indoor use are made from Propafilm C (Imperial Chemical Industries, Wilmington, North Carolina), which is relatively impervious to gaseous diffusion in very thin films, is transparent in the infrared, and easily heat sealed. This material is too fragile for very large canopies or in the field where stress from wind can be expected; here we use acrylic sheet glazing material (4 mm). This material can be glued and, when supported by a frame, forms large, durable canopies.

B. Environmental Control

Control of the aboveground atmosphere and temperature within the canopy is essentially similar to the methods described elsewhere. $^{14}CO_2$ may be generated as required by the addition of acid to $Na_2^{14}CO_3$ within the labeling apparatus. Alternatively, it may be advantageous to prepare $^{14}CO_2$ at the required specific activity in advance and trap the gas cryogenically in a small cylinder. The apparatus comprises a manifold with connections to: (1) an empty 2-lb CO_2 cylinder; (2) a Buchner flask containing $Na_2^{14}CO_3$ solution and fitted with separatory funnel containing H_2SO_4 (1 M); and (3) a vacuum pump and gauge. The manifold has an inline $CaCl_2$ trap for water. The system is evacuated, the vacuum pump is then isolated, and the CO_2 cylinder cooled by partial immersion in liquid N_2. $^{14}CO_2$ is generated by addition of acid such that the pressure within the manifold remains somewhat less than atmospheric. The $^{14}CO_2$ is trapped as a solid within the cylinder. Once the $^{14}CO_2$ collection is complete the cylinder valve is closed and the temperature allowed to return to ambient.

A valve is used to add the gas directly to the canopy atmosphere to maintain the required concentration and specific activity throughout the tracer pulse. The valve is computer controlled using digitized signals from an infrared gas analyser (IRGA) or rate meter equipped with a thin-window Geiger – Müller tube (see Chapter 2, this volume). Each demand for CO_2 results in the addition of a preset volume of gas from the storage cylinder followed by a latent period in which no addition is allowed; this damps oscillation in the control system. The amount and rate of $^{14}CO_2$ addition is derived from a record of the operation of the addition valve. This simplifies the handling of the $^{14}CO_2$ during the experiment and may provide better control of $^{14}CO_2$ addition than direct generation from carbonate within the apparatus.

It is important to obtain good control of soil moisture in cylinder experiments because of its effects on plant C partitioning and symbiont or soil microbial activities. Plant transpiration transfers water from the belowground to the aboveground canopy, where it condenses on the cooling coils. It may be necessary to include a drainage system to remove this water. Return of the condensate to the soil should be avoided because of its dissolved $^{14}CO_2$ content. Soil water loss is usually replaced by watering to weight at frequent intervals. In experiments with small cylinders the primary problem is the frequency at which water content must be adjusted. Large cylinders are difficult to move onto a suitable scale. In both cases such manipulations disrupt the apparatus used to contain and monitor the aboveground and belowground atmospheres. It is also difficult to maintain synchrony in wetting and drying cycles in different cylinders. Electronic bathroom scales that detect weight using strain gauges can be modified to be read by a computer. Cylinders of up to 150 kg can be permanently mounted on the scales. It then becomes possible to continuously monitor water content and either to maintain soil water within narrow limits or to control wetting and drying cycles by additions of water under programmed control. Tensiometers or conductivity measurements using Bouyoucos blocks can also be used. We prefer gravimetric methods; they are more reliable and are not restricted in working range.

C. $^{14}CO_2$ Pulse

The specific activity of the applied $^{14}CO_2$ and the length of the tracer pulse can be varied to obtain sufficient label in the pools of interest. A simple model of the anticipated ^{14}C distribution is useful in predicting tracer dilution and turnover times in various pools so that experiments can be designed to be sensitive to changes in the most dilute and slow pools of interest and approximate detection limits calculated. In general, investigations of C allocation to root systems and microbial symbionts require relatively large amounts of label to be assimilated because the tracer must pass through several intermediate compartments before reaching the symbiotic components. The "shape" of the ^{14}C pulse becomes broader and shallower in each successive C pool and may be difficult to detect, either in pools remote from the origin or in those with slow turnover, if insufficient tracer is incorporated. Short pulses at very high specific activity should yield the highest resolution of tracer dynamics because the shape of the peak of tracer concentration (specific activity) in dilute pools will be more easily discerned than with a broader pulse at lower specific activity. However, there are several other factors that do not favor the use of short pulses:

1. The possibility of radiation damage to plant leaf tissues, components of which may initially achieve specific activities close to that of the applied tracer. Sauerbeck and Führ (1963) found effects on growth in plants continuously exposed to $^{14}CO_2$ at a specific activity of 3.7 MBq/g C but not at 0.37 MBq/g C. This limit can be exceeded in pulse label experiments because of the transient nature of the radiation dose. We have used specific activities of 3.7–37 MBq/g C for short (1–2-hr) pulses without obvious effects on plant growth.

2. The transitions between $^{12}CO_2$ and $^{14}CO_2$ in the aboveground chamber cannot be instantaneous. In short pulses (<2 hr), the periods during which input tracer concentration is changing constitute an important part of the overall pulse and may greatly complicate subsequent interpretation.

3. The rebuilding of leaf photosynthate pools early in the photoperiod after nighttime depletion is likely to alter the source–sink balance such that belowground C allocation is reduced early in the photoperiod. A useful strategy for integrating diurnal variation in C allocation is to apply the label for one full photoperiod.

D. Belowground Respiration

The measurement of belowground respiration is important for several reasons. Respired CO_2 comprises an important component in the calculation of the C balance of all plants. This is particularly true of plants supporting microbial symbionts where up to 40% of the total C flux through the plant can be released as CO_2 belowground. Additionally, the respired $^{14}CO_2$ provides a valuable insight into the dynamics of tracer movement in the plant–symbiont system, which could otherwise be obtained only by sequential destructive harvest of many replicate plants or by the use of relatively expensive and inaccessible ^{11}C tracer techniques.

1. Methods of Measurement

The belowground evolution of CO_2 and $^{14}CO_2$ has been measured by two methods that may be classed as point sampling and total collection approaches. Point sampling has been applied particularly to plants labeled *in situ* (Warembourg and Paul, 1977). The method relies on the collection of small gas samples from tubes or sampling ports embedded at various positions in the rooting volume of the experimental plants. Samples withdrawn periodically for analysis are assumed to be in equilibrium with the soil atmosphere at the collection point. Such samples can give excellent information on the dynamics of the tracer pulse in the belowground atmosphere. However, derivation of quantitative estimates of total $^{14}CO_2$ from these data for C balance calculations requires an accurate model of diffusion in the soil atmosphere. It also involves uncertainty both in the volume of soil atmosphere sampled and in the distribution of the samples in relation to points of biological activity.

Where C balance data are required and the root–symbiont system can be at least enclosed in an open-bottomed container, it is preferable to collect or measure all CO_2 released in the belowground system. This has the advantage that the recovery of $^{14}CO_2$ will be essentially complete and avoids some of the uncertainties of the point collection approach. In the case of an open-bottomed, *in situ* cylinder, CO_2 is continuously purged from a headspace to be replaced by CO_2 diffusion from depth. Gas exchange will also occur through the embedded cylinder base; this must be estimated from a model of diffusion. The dynamics of the $^{14}CO_2$ release (and loss through the cylinder base) will be confounded by the lengths of diffusion paths from sources at different depths in the soil column to the surface headspace. An alternative may be to estimate ^{14}C respired belowground by mass balance calculation:

$$\text{net } ^{14}C \text{ belowground respiration} = \text{net } ^{14}C \text{ assimilation}$$
$$- [^{14}C \text{ biomass} + {}^{14}C \text{ shoot respiration} + {}^{14}C \text{ soil}].$$

In a closed cylinder the soil atmosphere can be purged by continuous passage of a flushing gas through the soil volume. The temporal accuracy of the measurements of respiration depend on the absorptive capacity of the soil and soil solution for CO_2 and on the efficiency of flushing. Figure 1 shows a closed cylinder design with airspaces above and below the soil column, which is supported on a perforated PVC plate. The flushing gas (air or CO_2-free air) is introduced into the tailspace of the cylinder, passes through the soil volume, and exits via the headspace. The flushing gas will flow primarily along preferential pathways and there may be pockets of soil atmosphere not in equilibrium with the flushing gas. The purge efficiency can be tested by introducing $^{14}CO_2$ into tubes prepositioned through the soil column base and observing washout. Purge flow rates depend on the respiratory activity of the system, but rates of about 2 ml/min/kg soil will yield exit concentrations of 1–2% CO_2 in typical systems.

Several plant–symbiont systems will normally be exposed simultaneously; it is therefore necessary to implement parallel soil atmosphere purge systems for each cylinder. We have used flow systems based either on suction or pressure as driver which, while very similar in design and using identical components, differ in their responses to leaks. This difference can be important, particularly in extended experiments where leaks at some point seem inevitable.

In the pressure-driven system (Fig. 1) a pressurized manifold (approximately 500 Pa) supplies gas to the tailspace of each cylinder via a Teflon capillary (1 m $\times$ 0.1 mm). The capillaries form almost all the resistance to flow through the system and thus regulate the flow rates. The gas leaving the cylinder is dried, passed to a stream selection valve and either directly to an individual CO_2 collection tube or, via an electronic flow transducer and IRGA, to a common CO_2 trap. The stream selection valve, flow transducer, and IRGA are under computer control using analog-to-digital (A-to-D) converters for signal measurement and relays to convert logic signals for valve and instrument control. Each gas stream is selected and measured sequentially. The headspace in the pressure-driven system is at slightly greater than atmospheric pressure because of the resistance to flow through the subsequent measuring devices and thus tends to leak out. This is immediately obvious because the measured flow rate decreases and the leak can be located and repaired. The concentrations of CO_2 and $^{14}CO_2$ are unaffected because the real purge flow rate of the cylinder is unchanged by the leak. The suction-driven system has the capillary and manifold moved to the effluent side. The headspace pressure is less than atmospheric and thus tends to leak inwards. While this arrangement is slightly less susceptible to leaks because the pressure differential across the headspace seals is smaller than that of the pressure-driven system, a leak is not signaled by any obvious change in system behavior. The resulting contamination of the effluent stream by aboveground or canopy atmosphere may lead to undetected errors in the data.

2. Specific Activity of Belowground Respiration

In a pulse-chase experiment the specific activity of the CO_2 respired by the root – symbiont – rhizosphere biota will show a general form similar to that of the curves in Fig. 2. Specific activity reaches a maximum, then decays to an asymptote after several days. Curves may be complex, with subsidiary local maxima, when source pools diurnally alternate between sucrose and stored starch reserves. At the asymptote it is assumed that the labeled C has achieved a stable distribution within the plant – symbiont biomass and that further release results from maintenance and turnover rather than growth. Destructive harvest and analysis of the plant, microbial, and soil components for ^{14}C at this point can be used to determine the C allocation pattern of the

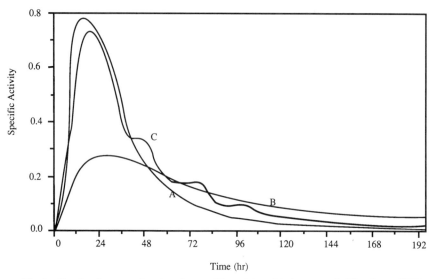

Fig. 2 Simulated curves showing the specific activity of belowground CO_2 after a $^{14}CO_2$ pulse from 0 to 16 hr in soybean plants: (A) with VA mycorrhizal and rhizobial symbioses; (B) no symbionts; and (C) with symbionts when starch reserves are remobilized at night.

system. Summation of ^{14}C released as $^{14}CO_2$ throughout the experiment completes the C balance:

$$\text{net photosynthesis} = {}^{14}C \text{ in biomass (plant + symbiont + soil microbes)}$$
$$+ {}^{14}C \text{ respiration.}$$

It is often possible to detect two or three components with different rate constants within the decay curve of belowground $^{14}CO_2$ respiration by curve peeling using nonlinear regression. In principle, these rate constants describe measurable pools (e.g., root nodule sucrose) and could thus be used to differentiate respiratory activity in different biological components of symbiotic root systems. In practice, any realistic compartmental model of root–symbiont C flow is much too complex to be susceptible to analysis by respiratory data alone. Measurements of the specific activities of many intermediate and precursor pools, which are currently not well understood for many systems even qualitatively, would be required.

Respiration of individual symbiotic partners can be estimated by between-treatment comparisons of total belowground $^{14}CO_2$. Calculations are based on specific rates of respiration ($^{14}CO_2$/g per unit tracer pulse length) to enable root systems of differing masses to be compared:

$$\text{symbiotic respiration} = \text{total respiration}$$

$$- \left\{ \left[\frac{\text{nonsym. root resp.}}{\text{nonsym. root dw}} \right] \times [\text{sym. root} - \text{symbiont dw}] \right\}$$

This calculation involves a number of simplifying assumptions:

1. Respiration in root–symbiont associations is strictly additive, that is, the presence of a microbial symbiont does not affect the respiration of the host root or of the native soil organisms.
2. All $^{14}CO_2$ released in a cylinder containing a root system without symbiotic associates is from plant root respiration plus nonsymbiotic native soil organisms.
3. $^{14}CO_2$ once released is not recycled by carboxylase enzymes to a significant extent.

While it is true that because of the above simplifications this calculation cannot accurately determine microbial symbiont respiration per se, it does represent the overall cost of the symbiosis in terms of additional belowground respiration; this information is equally useful.

E. Measurement of ^{14}C in Plant and Microbial Biomass

Labeled C assimilated during the applied pulse of $^{14}CO_2$ will eventually achieve a relatively stable distribution in structural and long-term storage components of plants and microbes and reallocation and metabolism of labeled C within the plant–symbiont system will be slow. This is revealed by the asymptote of the release of $^{14}CO_2$ in belowground respiration (Fig. 2). Analysis of the ^{14}C content of components such as roots, nodules, and VA mycorrhizal fungi at this point will reveal the C-allocation pattern of the system. Physical and chemical fractionation of plant, soil, and microbes can be performed to the required level of detail. We are here most concerned with allocation within the root–microbe symbiosis, but allocation between plant organs and measurements of C uptake rates by leaves are also of interest.

At harvest it is desirable to stop biological activity as rapidly as possible. Plant shoot organs should be excised, any measurements such as leaf area, requiring fresh tissue, should be made immediately, and the tissues transferred to a forced air oven for rapid drying. For measurements of labile biochemical components, subsamples frozen in liquid nitrogen and freeze dried may be preferable. It is much more difficult to rapidly harvest the belowground system. Activity can be slowed by cooling the cylinders to 4°C

Table 1
^{14}C Allocation in a Soybean–*Rhizobium*–*Glomus* symbiosis

Week	6		9	
	C(mg)	%	C(mg)	%
Shoot	67.2	50.7	127.2	61.2
Shoot respiration	8.4	6.3	8.2	3.9
Roots	10.4	7.8	16.9	8.1
Nodules	2.6	2.0	3.6	1.7
Mycorrhizal fungus	3.6	2.7	5.8	2.8
Root washings (+) soil[1]	2.4	1.8	2.7	1.3
Belowground respiration				
Roots + soil	6.9	5.2	13.5	6.5
Nodules[2]	12.4	9.3	20.4	9.8
Mycorriza[2]	18.2	13.7	9.7	4.7
Total	132.5		207.8	

Source: Adapted from Harris *et al.* (1985).
[1] Includes exudates and microbial biomass.
[2] Calculated by difference between symbiotic and nonsymbiotic plants.

but a complete dissection of the soil–root system may take several hours. It is seldom practical to completely separate roots from the whole soil column and some form of subsampling must be used. The sampling scheme should take account of the distribution of roots within the soil column.

Root nodules can easily be collected by excision. Intraradicle mycorrhizal fungal biomass can be estimated by morphometric methods (Toth and Toth, 1982) or by chemical estimation of chitin. The chemical method (Pacovsky and Bethlenfalvay, 1982) is preferred because it is quicker and because the hydrolysis products of chitin (*n*-acetylglucosamines) are isolated; these can then be assayed for ^{14}C. The biomass of extraradicle mycorrhizal fungus is difficult to measure in natural soils. Reliable measurements await the development of a suitable method for recognizing or separating mycorrhizal hyphae from the general soil fungal population. In soil-free systems the ratio of intraradicle to extraradicle fungal tissue is approximately unity in young soybean plants but increases with age. In natural soils it is probable that the extraradicle component will be reduced by grazing and is therefore overestimated by comparison to soil-free cultures. The ^{14}C content of extraradicle hyphae and chlamydospores can be measured directly after picking material from root surfaces using fine forceps and a dissecting microscope. The ^{14}C distribution in a tripartite symbiosis is shown in Table 1 as an example C budget for a symbiotic system.

III. COMMENTS

Calculations of C allocation and respiration in plant – symbiont systems based on comparisons between symbiont-infected and uninfected control plants assume that the control plant is physiologically similar to the infected plant except for the additional activity of the microbial symbiont. It is possible to avoid "big plant versus little plant" comparisons by adding appropriate nutrients to the uninfected plants such that all are of similar biomass at the time of the pulse label experiment. However, tissue nutrient concentrations, instantaneous growth rates, and other physiological parameters will vary between treatments, confounding measurements of the activity of the symbionts. This problem is inherently unavoidable in such comparisons whenever the presence of the symbiont has physiological effects.

One possible approach that avoids some of this difficulty is to arrange for physical compartmentalization of the symbiosis within a single host. This can be achieved, for example, in split-root systems where the symbiont(s) are confined to a portion of the root system but both infected and uninfected root portions are supplied by the same shoot. Pulse-chase label experiments are "snapshots" of C uptake and allocation at the time of the pulse. Single experiments cannot therefore illustrate the development and decay of symbiont activities, and several sequential experiments are needed. An alternative approach is to combine fewer pulse-chase experiments with conventional growth analysis and continuous measurements of belowground respiration.

ACKNOWLEDGMENTS

These studies were supported by NSF – LTER Project No. BSR 8702332 and NSF – CME Project No. DIR 8809640.

REFERENCES

Atkins, C. A., Herridge, D. F., and Pate, J. S. (1978). The economy of carbon and nitrogen in nitrogen-fixing annual legumes: Experimental observations and theoretical considerations. pp. 211 – 240. In "Isotopes in Biological Dinitrogen Fixation." International Atomic Energy Agency, Vienna.

Harris, D., Pacovsky, R. S., and Paul, E. A. (1985). Carbon economy of soybean – *Rhizobium-Glomus* associations. *New Phytol.* **101**, 427 – 440.

Harris, D., and Paul, E. A. (1987). Carbon requirements of vesicular-arbuscular mycorrhizae. In "Ecophysiology of VA mycorrhizal plants" (G. R. Safir, ed.) pp. 93 – 106. CRC Press, Boca Raton, Florida.

Pacovsky, R. S., and Bethlenfalvay, G. J. (1982). Measurement of the extraradicle mycelium of a vesicular-arbuscular mycorrhizal fungus in soil by chitin determination. *Plant and Soil* **68,** 143–147.

Sauerbeck, D., and Führ, F. (1963). Experiences on labelling whole plants with carbon-14. pp. 391–398. In FAO/IEAE, "The use of isotopes in soil organic matter studies." Brunswick Volkenrode.

Toth, R., and Toth, D. (1982). Quantifying vesicular-arbuscular mycorrhizae using a morphometric technique. *Mycologia* **74,** 182–187.

Warembourg, F. R., and Paul, E. A. (1977). Seasonal transfers of assimilated ^{14}C in grassland: plant production and turnover, soil and plant respiration. *Soil Biol. Biochem.* **9,** 295–301.

4

Photosynthesis/Translocation: Aquatic

A. Marshall Pregnall

Biology Department
Vassar College
Poughkeepsie, New York 12601

I. INTRODUCTION

Some of the earliest scientific use of ^{14}C was in the study and measurement of photosynthesis and plant production (e.g., Steeman Nielsen, 1952). The high sensitivity with which ^{14}C can be localized and quantified matches the rapidity with which plant cells assimilate and transform carbon. Thus, use of this isotope pervades the study of plant physiology and production. The chief theoretical drawback to use of ^{14}C for the study of production is the difficulty or ambiguity of measuring cell respiration relative to carbon assimilation; that is, ^{14}C assimilation probably measures something in between net production and gross production.

Partly owing to this difficulty, but more frequently owing to the greater cost and precautions for using isotope, aquatic production studies often use a variety of alternative methods. The most widespread of these is the oxygen change, or light-bottle dark-bottle method, which determines net photosynthesis in transparent bottles and respiration in darkened bottles. The change in oxygen with time may be followed as endpoint differences from Winkler titrations of dissolved oxygen content or continuously by recording oxygen electrodes. The technique is simple, inexpensive, and reproducible. However, while one can measure net photosynthesis and respiration, and thus gross photosynthesis, it must be remembered that the changes are due to activities of the entire plankton community rather than just those of autotrophs. In waters with low densities of phytoplankton, the magnitude of the dissolved oxygen changes may be small.

Another technique monitors the total dissolved inorganic carbon changes from photosynthesis and respiration as they are reflected in pH changes. One must know how much of the total alkalinity is comprised by inorganic carbon in order to calculate carbon assimilation. A modification of this principle for aerial measurements involves following gaseous CO_2 changes by infrared gas analysis (Holmes and Mahall, 1982). One can determine respiratory loss of CO_2, which is advantageous, but the sensitivity of the procedure for measuring carbon uptake is less than that when using ^{14}C.

If one is more interested in long-term net assimilation than in short-term production rates, one can monitor the changes in biomass through time using chlorophyll concentrations (as determined by extractions of different samples or repeated measurement of fluorescence yield of the same sample) for phytoplankton or dry weight and ash-free dry weight per unit area for macrophytes. Such measures usually ignore the biomass that was produced but has subsequently been released as dissolved organic material, consumed by herbivores or exported out of the system of interest, and thus provide underestimates of actual production.

The extreme sensitivity with which one can detect ^{14}C assimilation and the very short incubation periods required to obtain levels of radioactivity that can be differentiated between treatments have led to widespread use of the radiocarbon method. As techniques and materials for sample processing and scintillation counting improve, it is becoming much easier to perform the many correction steps that are required for calculating actual carbon assimilation from isotope incorporation.

II. SOURCES OF ^{14}C FOR PRODUCTION MEASUREMENTS

In order to perform measurements of production using ^{14}C, one must first decide what level of activity is desired and in what form the isotope will be added to phytoplankton or plant material during the incubation. Fortunately, a number of convenient options are available. The most widely used form for production studies is $NaH^{14}CO_3$, which is available commercially in predispensed ampoules with sterile aqueous or sterile dilute ethanol solutions of 5, 10, and 20 μCi (185, 370, and 740 kBq) per ampoule. These are easily stored, transported, and used in both lab and field. If necessary, several ampoules can be opened ahead of time and mixed in a larger volume to permit more rapid addition of isotope to many replicate vessels and ensure homogeneous distribution of activity among the vessels. $NaH^{14}CO_3$ is also available in crystalline form, which can be made up to any desired activity in solution prior to use. It can then be dispensed into ampoules, sealed, and sterilized for longer storage or used immediately for replicate additions. $Ba^{14}CO_3$ is also available in crystalline form and may be made up to desired activity in solution and used in the same manner as that for crystalline $NaH^{14}CO_3$ above. Alternatively, $^{14}CO_2$ can be evolved by acidifying $Ba^{14}CO_3$ and trapped in an alkaline solution for use. There are concerns regarding the presence of dissolved metals and other potentially inhibitory or stimulatory substances in solutions that have been stored in ampoules for long periods, and it is recommended that researchers working in oceanic or oligotrophic waters clean the isotope solutions with chelating resins prior to incubation (Carpenter and Lively, 1980).

III. SAMPLE PROTOCOL FOR MEASURING PHYTOPLANKTON PRODUCTION

Let us consider an idealized study of phytoplankton production at depth intervals through the water column in a lake or coastal area. Assume that a protocol has been filed with the campus Radiation Safety Officer and with the Nuclear Regulatory Commission (NRC) to transport isotope off campus and to use isotope outside; permission has been received from both the NRC and from local landowners to use isotope and perform our experiments at the chosen site. The experiment has been planned well ahead of time, and we have come to the site prepared with more than enough incubation vessels, filters, isotope ampoules, or solution, and spares of all the lab gear that we will use, such as sample bottles, pipets and pipet tips, filters, filter forceps, blotting paper for our working surface, and gloves. We have previously determined the depth intervals from which samples will be collected and have prepared a line with small clips fastened at each of the depths of interest. Bottles will be attached to the clips, and the whole array will be suspended from a large float. We also have a light meter and oxygen/temperature meter for measuring the conditions that cells at different depths experience. Figure 1 shows a generalized flowsheet for the experiment.

A. Collection Procedure

At each depth of interest, three types of samples need to be collected.

1. Phytoplankton samples for incubations
   - Starting at the surface and progressing downward, collect phytoplankton from predetermined depths using a water sampler such as a Van Dorn bottle, or a gentle peristaltic pump such as a hand-cranked

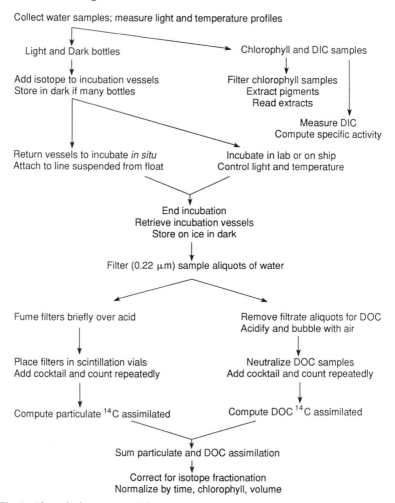

Fig. 1 Phytoplankton protocol flowsheet.

"jackrabbit" pump with weighted tubing lowered to the depths of interest.

- Fill the incubation vessels so that they overflow for several volumes in order to rinse the vessel and thoroughly displace any air bubbles that might alter oxygen concentration during the incubation.
- Minimize exposure of phytoplankton in water from depth to the higher irradiances and temperatures at the surface by storing the filled incubation vessels in a cool dark chest, such as a cooler with a little bit of ice, until all samples have been collected and are ready to receive isotope.

2. Samples for chlorophyll determinations
   • Fill a darkened 1-l sample bottle for subsequent determination of chlorophyll.
   • Store in the dark on ice until the sample can be filtered.
3. Samples for dissolved inorganic carbon (DIC) determination
   • Fill a darkened 1-l sample bottle for subsequent determination of DIC.
   • Measure and record the temperature of the water as it is being pumped into the sample bottle.
   • Store in the dark on ice until the sample can be filtered.

B. Incubation Vessels

The incubation vessels must be sufficiently large to contain enough phytoplankton to assimilate detectable amounts of isotope.

1. Glass 300-ml biochemical oxygen demand, (BOD) bottles
   BOD bottles are readily available, convenient to store and clean, contain a useful sample size, can be held on board ship in incubation tubs, incubated *in situ,* or incubated in the lab.
   • Add a glass marble to each bottle to improve water mixing (Strickland and Parsons, 1972).
   • Use at least one dark bottle and two light bottles for each depth; two dark and three light bottles would be better but may be impractical if many depths are being sampled.
   • If the bottles will be incubated while suspended from a line in the water column (see below), each bottle should have a loop of line tied securely around its neck that can be fastened to a clip on the line.
     a. Advantages: Relatively small size permits considerable sample replication. Simultaneous incubations of light and dark bottles for oxygen changes can be made using other BOD bottles, which eliminates confounding factors of different incubation vessel character.
     b. Disadvantages: There is evidence that a substantial "bottle effect" occurs in containers of 300 ml or less that may not occur with larger vessels (see Carpenter and Lively, 1980).
2. Plexiglas cylinders
   Cylinders are useful for incubations on the deck of a ship or in a lab. Tubes can be fitted with serum stoppers or ports through which to inject isotope and remove water samples periodically. Each cylinder can be surrounded by a thin water jacket (another plexiglas cylinder of slightly larger diameter) to control temperature individually.
   • Add a marble to enhance mixing in the tube.

- Use neutral density screening to provide a range of light levels appropriate for the depths from which water samples were collected.

C. Provision of Isotope

One can either use one ampoule of isotope per incubation vessel or prepare a stock solution of isotope ahead of time by opening many ampoules and emptying them into a small flask or bottle. The desired volume of isotope solution can then be added very quickly to many incubation vessels by repeatedly pipetting out of the flask or bottle. Great care should be taken not to spill the contents of the bottle, which may be stabilized by placing it in a flat-bottomed vessel such as a beaker or in a small board with a sufficiently large hole in it.

1. Work over absorbent blotting material to trap stray droplets or spills of isotope.
    - In the field, use heavy trays with blotting paper taped in place in order to keep things from blowing away.
    - In small boats, work low to the deck in order to keep things from pitching over suddenly.
    - Keep a small isotope waste container nearby for disposal of empty ampoules, pipet tips, wipes, etc.
2. Pipet aliquots of a prepared stock isotope solution directly into the bottom of each incubation vessel.
3. Open individual ampoules of isotope carefully.
    - Tap all the isotope solution into the bottom of the ampoule.
    - Etch the narrowest part of the neck using a small triangular file. It usually is not necessary to score the neck all around; a noticeable groove in one side will do.
    - Surround the ampoule with a lab wipe and gently *pull* the top off with slight pressure to one side; it is not necessary to snap the ampoule in two with pressure at the neck. If the top does not pull off easily, use the file to etch a bit more and try again.
    - Draw up the isotope solution as completely as possible from the ampoule into a syringe or pipet and dispense it into the bottom of the incubation vessel.
    - Rinse the ampoule with 1 or 2 ml of water, draw the rinse up in the same syringe or pipet, and add the rinse to the vessel.

D. Incubation

1. Incubations *in situ:*
    a. For work in a shallow pond:

- Place the bottles into a tray and lower to the bottom of the pond or place in shallow water with a small float to facilitate retrieval.

    b. For incubations at depth intervals in the water column:
- Start with the greatest depth and attach the bottles to clips on a weighted line, which should be spaced at the predetermined collection intervals.
- Suspend the line from a sturdy float or small boat, with the line and the surface bottles held out to the side in some manner to avoid potential shading by the boat.
- Monitor the incubation. Although it may be tempting to leave the area for the next few hours, it is required that field experiments involving radioactive isotopes be attended at all times. More to the point, it is unlikely that any abandoned float or boat will go uninspected by passersby during a period of hours; for their safety and the unmolested completion of your experiment, stay at the study site.

2. Incubations on a ship:
- Place the incubation vessels in a temperature-controlled bath or baths to mimic the ambient temperature range experienced by the samples *in situ* and control irradiance over the bottles with neutral-density screening.
- If desired, sample the bottles or plexiglas cylinders at intervals during the incubation.

3. Incubations in a laboratory:

The chief difficulties for laboratory incubations are providing enough light while simultaneously regulating the temperature to prevent overheating.
- Place the incubation vessels in a large tub connected to a recirculating water bath. If samples are from different depths with very different temperatures, several tubs with discrete temperature control will be necessary.
- Use different amounts of neutral-density screening to simulate light attenuation through the water column.
- Suspend large floodlamps over the tubs in which incubation vessels are placed.

E. Additional Measurements

Chlorophyll concentration and DIC concentration must also be determined in order to determine the specific activity of the isotope in the incubation vessels and to normalize the fixation of isotope per unit of biomass in the incubation vessels. Use the additional water samples that were set aside during the initial water collections.

- Chlorophyll concentration of water from different depths, and perhaps phytoplankton cell numbers at those depths, can be determined on water samples using procedures described elsewhere (Strickland and Parsons, 1972).
- Total DIC content of water at the different depths must be known in order to compute the specific activity of the isotope in the incubation vessels. The measurement of carbonate alkalinity is usually a straightforward procedure, unless additional ions contribute substantially to total alkalinity (Strickland and Parsons, 1972; Gieskes and Rogers, 1973).
- It is useful if not essential to know the light and temperature regimes of the sample area.

F. Stopping Incubations

- Retrieve the incubation vessels and place them in the dark on ice if there will be a delay until further processing, for example, for incubations *in situ*.
- Avoid using fixative of any sort (dilute formalin, gluteraldehyde, etc.), if considerable leakage of assimilated carbon occurs (Carpenter and Lively, 1980).
- It is recommended that 0.22-μm pore filters be used unless one intentionally wants to fractionate the plankton cells by size. While a variety of nominal pore sizes have been used (0.22, 0.45, and 0.8 μm), it has been demonstrated repeatedly that small autotrophs such as cyanobacteria can pass readily through the larger pore-size filters. Small diameter (25 mm) filters will fit conveniently into the bottom of scintillation vials, but have lower filtration volume capacity than larger (47 mm) filters. Some membrane filters are soluble in xylene- or toluene-based scintillation cocktails, while others are not.
- Briefly rinse the filter both before and after sample filtration with small volumes of previously filtered water, initially to saturate the filter with water, and subsequently to help remove unincorporated isotope from the filter.
- Filter aliquots (10 to 300 ml, depending upon the density of phytoplankton and other particulate matter) of the incubated samples using low vacuum (e.g., <250 mm Hg) so as not to rupture or shear cells, which would release assimilated carbon.
- Fume filters *briefly* (1 to 10 min) over acid to drive off excess $^{14}CO_2$ that may have been retained in the filters. This can be done in a fume hood with an old desiccator; place a dish of concentrated HCl or HNO_3 in the base, replace the inner shelf, then lay the filters gently on the shelf for a few minutes. It is essential that you keep track of the sample origin of the

filters while they are in the desiccator. Caution should also be exercised while maneuvering filters in a fume hood so that they do not get sucked up the hood chimney owing to the force of the fan.

G. Measurement of Dissolved Organic Carbon (DOC)

Since it is now widely acknowledged that healthy phytoplankton cells release assimilated carbon (Bjørnsen, 1988), we must make sure to determine how much assimilated activity has been excreted into the medium over the course of the incubation. Additionally, potential contamination of stock isotope solution by acid-stable, nonbicarbonate compounds should be assessed to ensure that we do not mistakenly attribute counts from such contaminants to algal release.

1. To detect acid-stable contaminants:
   - Add an aliquot of stock isotope solution to a volume of filtered water equivalent to that used for the actual incubations.
   - Immediately remove several replicate 5- to 10-ml samples and place them in scintillation vials.
   - Add 1 ml of 5.5N HCl and air bubble for 15 to 20 min in a fume hood to drive off $^{14}CO_2$.
   - Neutralize with an equivalent amount of NaOH and add scintillation cocktail that will mix with aqueous samples.
   - Owing to the presence of water, salts, and possible excesses of acid or base, spontaneous chemoluminescence will occur for some time. Count the vials repeatedly over a period of hours to days in order to determine how much stable activity is in the vials.
2. Algal release of DOC:
   - Measure either a time course of acid-stable activity during the incubation or an end point of acid-stable activity (Pregnall, 1983).
   - Remove aliquots (5–10 ml) of sample filtrate (0.22-μm pore size, filtered at low vacuum to remove particulates) at intervals during the incubation or at the end.
   - Acidify the filtrate with 1 ml 5.5N HCl and air bubble in a fume hood to drive off $^{14}CO_2$.
   - Neutralize with an equivalent amount of NaOH and add scintillation cocktail.
   - Count repeatedly until stable counts are observed. These counts should be corrected for any counts found in the stock isotope solution from above and for the volume of sample filtered. The corrected DOC

counts are added to the volume-corrected counts of the sample filters in order to obtain total activity assimilated.

H. Counting

Scintillation counting must be performed on the accumulated filters and DOC samples as well as DOC samples from the isotope stock solution. Quench correction curves are necessary for each method of sample preparation.

1. Quench correction
    a. Filters with phytoplankton:
 • Filter a range of volumes of unlabeled phytoplankton onto between 5 and 10 filters of the type used during the experiment, and allow the filters to dry briefly. The varying amounts of phytoplankton on the filters will provide a range of both color quench and the confounding pigment chemiluminescence.
 • Add a known activity of some nonvolatile isotope compound such as ^{14}C-sucrose to the filter and dry briefly.
 • Add the appropriate scintillation cocktail and count repeatedly over a period of hours to several days. Counting efficiency (CPM observed ÷ DPM of standard added) can be plotted for the different filters as a function of the volume of sample filtered, an external standards ratio, or a channels ratio value.
    b. Aqueous DOC samples:
 • Place a volume of unlabeled filtrate equivalent to the samples used for labeled DOC measurements into a scintillation vial.
 • Perform acidification, bubbling, and neutralization steps as above.
 • Add a known activity of the nonvolatile standard as for the filters above.
 • Add scintillation cocktail and count periodically until stable.

I. Sample Calculations

Let's consider an incubation using 300-ml BOD bottles to which 370 kBq (10 μCi) of $NaH^{14}CO_3$ has been added. The bottles were incubated *in situ* for 3 hr. Duplicate 10-ml aliquots were sampled for DOC release, and 30-ml samples were filtered for measurement of assimilated carbon. Total inorganic carbon was determined to be 3 mM (3000 μmol/l), and chlorophyll was 17 μg/l. The counting efficiency of filters was determined to be 0.78, while the counting efficiency of aqueous samples was 0.57.

1. Determine the specific activity of isotope added to the incubation, corrected for DIC content and incubation container volume

$$(3000 \ \mu\text{mol inorganic C/l} \times 0.3 \ \text{l}) \div 370 \ \text{kBq} = 2.43 \ \mu\text{mol C/kBq}$$

$$2.43 \ \mu\text{mol C/kBq} \div 6 \times 10^4 \ \text{DPM/kBq} = 4.054 \times 10^{-5} \ \mu\text{mol C/DPM}$$

2. Apply quench corrections to obtain the decays of the filter samples

$$\text{CPM} \div \text{counting efficiency} = \text{DPM};$$

thus, for a filter that had 7600 CPM

$$7600 \ \text{CPM} \div 0.78 = 9744 \ \text{DPM}$$

3. Correct for sample or aliquot fraction (i.e., 30 ml filtered for counting out of 300 ml incubated)

$$9744 \ \text{DPM/30 ml filtered} \times 10/10 = 97440 \ \text{DPM/300 ml incubated}$$

4. Convert activity incorporated by specific activity to obtain carbon units incorporated

$$97,440 \ \text{DPM/300 ml incubated} \times 4.054 \times 10^{-5} \mu\text{mol C/DPM}$$

$$= 3.950 \ \mu\text{mol C assimilated into particulate}$$

$$\text{matter in 300 ml during incubation}$$

5. Compute DOC release
   For example, 200 CPM in a 10-ml aliquot for DOC:

$$200 \ \text{CPM} \div 0.57 \ \text{counting efficiency} = 351 \ \text{DPM/10 ml}$$

$$351 \ \text{DPM/10 ml} \times 30/30 = 10,530 \ \text{DPM/300 ml incubated}$$

$$10,530 \ \text{DPM} \times 4.054 \times 10^{-5} \ \mu\text{mol C/DPM}$$

$$= 0.427 \ \mu\text{mol C in DOC in 300 ml}$$

6. Correct for discrimination against the heavier ^{14}C isotope during assimilation (add 5%) and add DOC to fixed carbon retained on filter to obtain total carbon fixation.

3.950 μmol $\times$ 1.05 (discrimination factor)

= 4.148 μmol C in particulate matter in 300 ml

0.427 μmol $\times$ 1.05 (discrimination factor)

= 0.448 μmol C in DOC in 300 ml

4.148 + 0.448

= 4.596 μmol C fixed in 300 ml, of which 9.8%

was released as DOC

7. Correct for chlorophyll content and duration of incubation to obtain carbon fixation rates in units of moles of carbon fixed per unit time per unit chlorophyll or volume

4.596 μmol C $\div$ 3 hr = 1.532 μmol C/hr in 300 ml

17 μg Chl/l $\times$ 0.3 l = 5.1 μg Chl in 300 ml

1.532 μmol C/hr in 300 ml $\div$ 5.1 μg Chl in 300 ml

= 0.300 μmol C/(μg Chl $\times$ hr)

or

1.532 μmol C/hr $\div$ 0.3 l = 5.107 μmol C/(l $\times$ hr)

8. Account for all the isotope: where has all the ^{14}C ended up?

	kBq	μmol C
Assimilated in cells	1.591	4.148
Released as DOC	0.185	0.448
Still in solution	368.224	895.000
Total	370.000	900.0

J. Waste Disposal

Any such experiment will generate labeled filters in scintillation vials and copious amounts of labeled water. Proper disposal of these wastes will depend on the quantity and activity of the material. Before performing any experiment, calculate what maximum activities will be in the water, assuming that all isotope remains in solution, and what maximum activity might be in the particulate matter, assuming that all isotope is fixed. Consult with the campus Radiation Safety Officer about proper disposal of your materials using such activities to guide your decisions.

IV. MODIFICATION OF THE PROTOCOL FOR OTHER PRODUCERS

The basic protocol outlined above can be modified to accomodate a wide variety of other primary producers from benthic microalgae to stony corals. Each producer will require its own adjustments to account for size or habitat, and one will always need to consider whether to choose specific parts or tissues versus the whole organism, the basis for tissue selection, the plant or tissue age, and the presence of epiphytes and potential effects of epiphyte removal.

A. Collection of Material

1. Benthic Microalgae

In soft-bottom communities, direct coring of sediments is the simplest way to obtain microalgal samples with minimum disturbance. The size of the cores will vary with the scale of interest; however, remember that larger cores have less edge perturbation relative to surface area exposed. Also remember that an entire microbial community is present in a sediment core, including photosynthetic prokaryotes and many heterotrophs, and that incubation of cores requires consideration of benthic respiration, potentially large differences in interstitial dissolved inorganic carbon over minute distances, and potential alteration of the dissolved inorganic carbon equilibria if one generates gaseous $^{14}CO_2$. Moreover, since gas-phase CO_2 concentrations are so low, depletion of CO_2 can occur rapidly.

In hard substratum areas, chip off parts of rock or collect entire cobbles as carefully as possible to avoid disturbing or dislodging the microalgal community. Alternatively, artificial hard substrata such as glass slides or porcelain tiles may be deployed at the field site well in advance of the labeling experiment in order to develop a microalgal community on a more readily managed substratum. If the collected substratum is to be returned to the laboratory for study, the samples should be kept moist or completely submerged to avoid desiccation stress.

2. Macroalgae

Decide whether entire plants or parts of plants (tips or fronds) will be used, and then regularize the selection of tissues with respect to age, epiphyte load, growth, and nutritional history (Arnold and Manley, 1985). As with phytoplankton, avoid exposure to high light and temperature for plants collected from dim light environments. Pay particular attention to epiphytes; removal prior to incubations will require consideration of potential injury during the removal or surface sterilization procedure.

3. Seagrasses

The considerations are generally the same as for macroalgae. The presence of gas-filled lacunae makes O_2-change measurements subject to error, and internal CO_2 recycling (Hartman and Brown, 1967) may confound carbon fixation determination. Consider leaf age and leaf turnover during tissue selection, as well as the leaf age- and season-dependent epiphyte load.

4. Emergent Plants (Marsh Plants, Mangroves, Emergent Aquatics, and Others)

These plants require more decisions regarding the choice of tissue sections versus entire plants, as well as the choice of aerial parts, submerged parts, or both. As with seagrasses, the presence of gas-filled lacunae and potential internal recycling of O_2 and CO_2 (Gleason and Zieman, 1981) must be considered when using gaseous provision of $^{14}CO_2$. Additionally, the small amount of gaseous CO_2 that is available relative to the large amounts of tissue that may be used should lead one to restrict aerial incubations to short duration.

5. Corals

Collection of corals is not too dissimilar from that of microalgae living on hard substratum. Minimize damage to the material while removing parts of the coral for placement in chambers; moreover, if an intact coral will be surrounded by a clear bag, take great care to avoid tearing the bag and releasing isotope to the surrounding waters. As with some macroalgae and most vascular plants, consider how selection of parts of morphologically differentiated corals will affect your interpretations.

B. Incubation Vessels

Macroalgae, seagrasses, and corals may require larger incubation vessels, such as large bottles, plexiglas chambers, and plastic bags. These may have fewer potential artifacts of bottle enclosure, such as wall effects, oxygen oversaturation, inorganic carbon or nutrient depletion, and temperature fluctuation than do the smaller BOD bottles. However, adequate mixing of water in such vessels is more difficult to achieve owing to the physical interference of the plant tissue; consequently, adequate mixing becomes more important. Additionally, more isotope may be required with the larger sample volumes, leading to the subsequent need for proper disposal of larger amounts of waste materials.

1. Plexiglas Chambers

Chambers can be made to many sizes and shapes, fitted with injection or sample withdrawal ports, and fitted with water-tight stirring motors and mixing bars or paddles. They can be incubated in the lab with good temperature control or replaced *in situ* (Brylinsky, 1977; Pregnall, 1983). Such chambers, or combined metal and plexiglas chambers, can also be used for studies of benthic microflora production while submerged or in air (Van Raalte *et al.*, 1974; Darley *et al.*, 1976).

2. Clear Plastic Bags

Large bags are useful for enclosing large plants or parts of plants *in situ*, for example, kelps or marsh plants (Towle and Pearse, 1973; Lobban, 1978). The bags can be fitted with injection ports and sample withdrawal ports. However, large amounts of labeled water and tissue are generated by these incubations, making bags more difficult to work with than smaller vessels. Whatever choice is made, ensure adequate water motion and adequate temperature control if not returned *in situ*. Additionally, if the light field of the sampled environment is to be mimicked, provide for light control if the vessel is not returned *in situ*.

C. Provision of Isotope

1. Incubations with bags, large bottles, or plexiglas chambers
   • Once the plant material has been placed in the incubation vessel, a volume of stock isotope solution may be added via pipet or injected through a serum stopper.
   • Ensure that vigorous mixing of incubation water occurs.
2. Aerial incubations of benthic microalgae or macrophytes
   • Once the plant material is securely in place, the incubation system is air tight, and the incubation can be started, gaseous $^{14}CO_2$ must be provided to the atmosphere of the incubation vessel. The isotope solution can be drawn up into a large-volume syringe (10–50 ml) and is acidified by drawing up a volume of acid (e.g., 85% lactic acid) into the syringe, followed by a relatively large volume of air. Gentle mixing of the syringe contents will release $^{14}CO_2$ into the air volume. Subsequently, the air alone is injected into the incubation vessel.

D. Additional Measurements

- As with phytoplankton protocols, the inorganic carbon that is available (dissolved or gaseous) and the chlorophyll content of the incubated tissue should be determined.
- The surface area and the fresh and dry weights of the macrophyte tissue or the surface area of microalgal exposure should be measured in order to normalize production to units of area and mass.

E. Stopping Incubations

- Remove the incubated material from the isotope solution and rinse any remaining isotope from the material as quickly as possible.
- When using large bags that have been replaced *in situ,* take great care to prevent loss of isotope to the surrounding area during retrieval.
- If plexiglas chambers have been used, remove the top of the chamber, remove the sample, and rinse it gently with cold filtered water.
- If there will be any extensive delay between retrieval of the incubation vessels and removal of the samples, place the vessels on ice in the dark to slow assimilation.
- After rinsing, macrophyte tissues should be quickly frozen or fixed in buffered formalin. While addition of fixatives to phytoplankton incubation vessels creates considerable difficulties (see above), removal of macrophyte tissues and fixing in small volumes of formalin will work if quick freezing is not practical. If fixative is used, additional sample processing will need to be done (see below).
- With corals, the polyp/symbiont tissue must be removed from the calcareous skeleton without losing too much of the assimilated carbon. One technique uses a dental Water Pik® to separate and collect the tissues (Johannes and Wiebe, 1970). If necessary, the cnidarian and algal symbiont tissues can be further separated by differential centrifugation. Separately or together, the tissues can subsequently be lyophilized or homogenized prior to scintillation counting.
- Since macrophytes also release organic carbon during photosynthesis (Wetzel, 1969; Brylinsky, 1977; Pregnall, 1983), one must sample the incubation medium as in the protocol for phytoplankton in order to include this component of production.

F. Counting

- Rinse macrophyte tissues thoroughly to remove remaining inorganic ^{14}C and then dry completely, either by lyophilization or in a warm oven. Some very volatile components may evaporate during warm drying, so lyophilization is preferred if readily available.

- Grind the dried tissues to a fine powder, then weigh out aliquots into scintillation vials.
- Add scintillation cocktail with water to create a stable gel with the tissue powder evenly distributed.
- If fixative was used to stop the incubations, count aliquots of fixative as well.
- Alternatively, digest the labeled tissues prior to mixing with scintillation cocktail. A variety of alkaline tissue solubilizers are available but may not be entirely effective on complex algal carbohydrates. Concentrated HNO_3 can also be used (Van Raalte *et al.,* 1974), but extended digestion may result in the loss of $^{14}CO_2$ unless it is trapped in alkali. Both very high and low pH may cause considerable spontaneous chemiluminescence during the counting procedure unless samples are neutralized.
- Construct quench correction curves for each sample preparation technique.

V. TRANSLOCATION STUDIES

In the morphologically complex macroalgae and vascular plants and in the symbiotic associations between algal cells and various heterotrophs, it is of great interest to determine how much of the carbon that is assimilated in the actively photosynthesizing cells or tissues gets translocated to the nonphotosynthetic tissues (roots, medullary tissue, polyps, etc.). It is also interesting to determine the structural and metabolic pathways of such translocation. While translocation studies using ^{14}C may be set up quite similarly to production studies, there are important differences in the incubation vessels, in how isotope is provided to the tissues, and in how the samples are subsequently processed. Figure 2 presents a flowsheet for a generalized translocation experiment using macrophytes.

A. Incubation Vessels

Since the intention of studying translocation is to determine the movement of material from one part of the organism to another, the incubation vessel will necessarily be modified in order to spatially separate the tissue that is being provided with isotope from the tissue that may be receiving translocated assimilation products. Two-chamber systems are used most often, in which a single vessel possesses a partition or seal through which some narrow part of the plant is passed, or in which two vessels contain the labeled versus unlabeled tissues and a bridge containing part of the plant provides a connection. Water-resistant lubricants such as lanolin are used to prevent the passage of incubation water from the labeled chamber to the

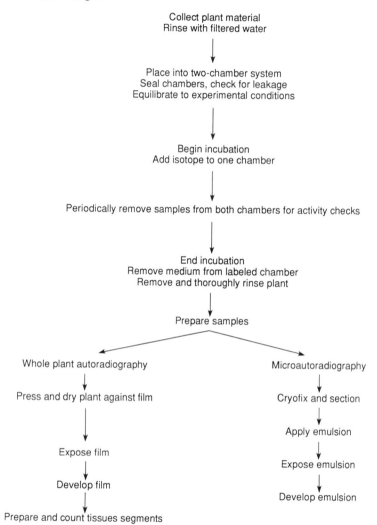

Fig. 2 Translocation protocol flowsheet.

unlabeled chamber. The integrity of the seal between chambers is crucial to the success of the experiment; however, manipulations or materials that may harm the plant's tissues must be avoided (see McRoy and Goering, 1974; Denny, 1980). After the plant is securely placed in the incubation system without leaks between chambers, equilibrate the system to the desired conditions of light and temperature.

B. Addition of Isotope

If one tissue or blade or frond is isolated in a two-chambered system, isotope may be added as for phytoplankton or macrophyte incubations (see above). However, if we wish to localize the provision of isotope to one spot on a frond, the point of application must be separately bagged or the entire frond kept moist during labeling (Lobban, 1978; Kirchman *et al.*, 1984).

C. Incubation

Translocation experiments usually last only a few hours, although labeling may continue for 24 hours in some cases.

- Maintain light and temperature conditions appropriately.
- Ensure that adequate mixing occurs in the labeled chamber.
- Ensure that the moist bridge does not dry out or that a seal between chambers is not disturbed.
- Remove water samples from both labeled and unlabeled chambers periodically for determination of potential isotope leakage through the seal or bridge and for determination of labeled DOC release.

D. Stopping Incubations

- Drain the isotope-containing chamber with a siphon or by aspiration into a vacuum flask in order to prevent isotope from contacting the unlabeled tissue.
- Remove the plant from the incubation system and rinse several times with filtered water in order to remove isotope on the surface.
- For quantitative determination of isotope translocation alone, separate the plant sample into parts and prepare them for scintillation counting as above. However, if we wish to determine the route of translocation as well as the amount, we must perform autoradiography on the plant prior to preparation for scintillation counting.

E. Localization of Translocation by Autoradiography

Autoradiography in translocation studies falls into two general categories (1) the use of films exposed against the whole plant; and (2) the use of emulsions coated onto tissue sections for histoautoradiography, or microautoradiography (see, e.g., Emerson *et al.*, 1982; Vogelmann and Dickson, 1982; Shih *et al.*, 1983).

1. Whole-plant autoradiograms
   - Press the rinsed plant flat and place it against a sheet of emission-sensitive film.
   - Quickly freeze or dry the preparation to prevent further metabolism of incorporated label. If this is done quickly, any shape change in the plant material from shrinkage or cracking should be small and will minimize distortion of the autoradiogram.
   - Expose the film for a period from days to many weeks depending upon the amount of isotope incorporated. Parallel incubations and determination of label incorporation will guide the choice of exposure time.
   - Following film exposure and development, one can see where label has moved through the plant, and consequently localize areas from which to obtain radioactive counts. Localized sections of the dried, pressed plant material can be ground to a powder for scintillation counting as described above.
2. Histoautoradiography

In histoautoradiography, or microautoradiography, tissue sections are prepared using cryohistological techniques, followed by application of decay-sensitive emulsion in order to localize translocation elements. Details of the tissue preparation vary depending upon the type of plant used and the eventual resolution required (see Emerson *et al.,* 1982; Vogelmann and Dickson, 1982; Shih *et al.,* 1983). Because of the lower radioactivity that may be present in a thin tissue section, exposure to the emulsion typically lasts longer than for whole-plant autoradiography.

VI. COMMENTS

A. Reminders

There are several things that must be determined during preliminary test incubations before one begins to perform lab or field incubations to collect data. Without such information, the eventual interpretation or extrapolation of data is unnecessarily restricted.

1. Determine the linearity of ^{14}C-activity incorporation with time. Endpoint determinations for incubations of longer than an hour or two can severely underestimate the total production that occurs.

2. Assess the possibility of isotope or total inorganic carbon depletion as an unforseen consequence of using too much plant material, incubating the vessels for too long, or having incubation water with low dissolved inorganic

carbon. Concurrently, check for excessive increases in dissolved oxygen during incubations, which would result from similar causes. Either situation will lower carbon assimilation through time. Carbon depletion can occur rapidly during incubations in air, forcing one to use short incubations and low amounts of tissue.

3. Consider checking for endogenous rhythms or fluctuations in carbon fixation capacity of the organism. A series of overlapping short incubations provide much more information than a single long incubation.

4. If possible, learn something about the nutrient and/or growth history of the sample material. There is considerable interaction between nutrient status, photosynthesis, and growth potential.

5. Check the isotope stock solution for purity and clean it of potentially inhibitory substances before use in oceanic and oligotrophic waters.

6. Check for DOC leakage from samples during the incubation.

7. Account for all isotope activity at the end of the experiment.

8. Use the cleanest equipment and materials possible.

9. When feasible, use larger containers; the size of the container affects the vigor and health of the experimental material, as well as setting the limits of nutrient and inorganic carbon availability.

B.   What Is Measured by ^{14}C Incorporation?

Despite the many thousands of production measurements that have been made using ^{14}C over nearly 40 years, it remains somewhat ambiguous as to what ^{14}C incorporation actually measures. In part, it depends upon the duration of incubation and the plant or algal type being used. For macrophytes, short incubations of only an hour or so in length should give ^{14}C assimilation rates that closely correspond to gross carbon fixation, for respiratory loss of $^{14}CO_2$ from internal metabolite pools lags behind photosynthetic fixation into those pools. For macrophytes in prolonged incubations of more than an hour, isotope incorporation approaches net carbon assimilation rates as internal carbon pools begin to equilibrate with isotope and respiratory losses occur. However, long incubations have the experimental artifacts of dissolved inorganic carbon and nutrient depletion as well as increasing oxygen in the light, all of which stem from enclosing the plant sample in a finite container. For phytoplankton cells and microalgae, isotopic equilibration can occur much more rapidly, so incubations longer than a few minutes may approach net carbon assimilation. During some experiment, try to perform simultaneous incubations using light and dark containers for the determination of oxygen production and consumption in order to compare between the two methods and compute carbon : oxygen quotients. The changes in oxygen may give some insight into your ^{14}C assimilation data.

REFERENCES

Arnold, K. E., and Manley, S. L. (1985). Carbon allocation in *Macrocyctis pyrifera* (Phaeophyta): intrinsic variability in photosynthesis and respiration. *J. Phycol.* **21**, 154–167.

Bjørnsen, P. K. (1988). Phytoplankton exudation of organic matter: *Why* do healthy cells do it? *Limnol. Oceanogr.* **33**, 151–154.

Brylinsky, M. (1977). Release of dissolved organic matter by some marine macrophytes. *Mar. Biol.* **39**, 213–220.

Carpenter, E. J., and Lively, J. S. (1980). Review of estimates of algal growth using ^{14}C tracer techniques. pp. 161–178. In "Primary Productivity in the Sea" (P. G. Falkowski, ed.). Plenum Press, New York.

Darley, W. M., Dunn, E. L., Holmes, K. S., and Larew, III, H. G. (1976). A ^{14}C method for measuring epibenthic microalgal productivity in air. *J. Exp. Mar. Biol. Ecol.* **25**, 207–217.

Denny, P. (1980). Solute movement in submerged angiosperms. *Biol. Rev.* **55**, 65–92.

Emerson, C. J., Buggeln, R. G., and Bal, A. K. (1982). Translocation in *Saccorhiza dermatodea* (Laminariales, Phaeophyceae): anatomy and physiology. *Can. J. Bot.* **60**, 2164–2184.

Gieskes, J. M., and Rogers, W. C. (1973). Alkalinity determination in interstitial waters of marine sediments. *J. Sed. Petrol.* **43**, 272–277.

Gleason, M. L., and Zieman, J. C. (1981). Influence of tidal inundation on internal oxygen supply of *Spartina alterniflora* and *Spartina patens*. *Estuar. Coast. Shelf Sci.* **13**, 47–57.

Hartman, R. T., and Brown, D. L. (1967). Changes in internal atmosphere of submersed vascular hydrophytes in relation to photosynthesis. *Ecology* **48**, 252–258.

Holmes, R. W., and Mahall, B. E. (1982). Preliminary observations on the effects of flooding and desiccation upon the net photosynthetic rates of high intertidal estuarine sediments. *Limnol. Oceanogr.* **27**, 954–958.

Johannes, R. E., and Wiebe, W. J. (1970). Method for determination of coral tissue biomass and composition. *Limnol. Oceanogr.* **15**, 822–824.

Kirchman, D. L., Mazzella, L., Alberte, R. S., and Mitchell, R. (1984). Epiphytic bacterial production on *Zostera marina*. *Mar. Ecol. Prog. Ser.* **15**, 117–123.

Lobban, C. S. (1978). Translocation of ^{14}C in *Macrocystis integrifolia* (Phaeophyceae). *J. Phycol.* **14**, 178–182.

McRoy, C. P., and Goering, J. J. (1974). Nutrient transfer between the seagrass *Zostera marina* and its epiphytes. *Nature (London)* **248**, 173–174.

Pregnall, A. M. (1983). Release of dissolved organic carbon from the estuarine intertidal macroalga *Enteromorpha prolifera*. *Mar. Biol.* **73**, 37–42.

Shih, M. L., Floch, J.-Y., and Srivastiva, L. M. (1983). Localization of ^{14}C-labeled assimilates in sieve elements of *Macrocystis integrifloia* by histoautoradiography. *Can. J. Bot.* **61**, 157–163.

Steeman Nielsen, E. (1952). The use of radioactive carbon (C_{14}) for measuring organic production in the sea. *J. Con. Int. Explor. Mer.* **18**, 117–140.

Strickland, J. D. H., and Parsons, T. R. (1972). A practical handbook of seawater analysis. *Bull. Fish. Res. Bd. Can.*, No. 167.

Towle, D. W., and Pearse, J. S. (1973). Production of the giant kelp, *Macrocystis*, estimated by *in situ* incorporation of ^{14}C in polyethylene bags. *Limnol. Oceanogr.* **18**, 155–159.

Van Raalte, C., Stewart, W. C., Valiela, I., and Carpenter, E. J. (1974). A ^{14}C technique for measuring algal productivity in salt marsh muds. *Bot. Mar.* **17,** 186–188.

Vogelmann, T. C., and Dickson, R. D. (1982). Microautoradiography of water-soluble compounds in plant tissue after freeze-drying and pressure infiltration with epoxy resin. *Plant Physiol.* **70,** 606–609.

Wetzel, R. G. (1969). Excretion of dissolved organic compounds by aquatic macrophytes. *BioScience* **19,** 539–540.

5

Microbe/Plant/Soil Interactions

R. P. Voroney

J. P. Winter

University of Guelph
Ontario Agricultural College
Department of Land Resource Science
Guelph, Ontario, Canada N1G 2W1

E. G. Gregorich

Land Resource Research Centre
Agriculture Canada
Central Experimental Farm
Ottawa, Ontario, Canada K1A 0C6

I. INTRODUCTION

[14]C-labeled organic compounds have been extensively used to study rates of decomposition (mineralization) and stabilization of microbial products (physical protection and humification). Substrates have ranged from simple, well-defined compounds such as glucose, acetate, and amino acids to more complex compounds such as cellulose, hemicellulose, and proteins, to plant leaves and roots, manure, or high-molecular-weight compounds. Experiments have been as short as a few minutes, and they have extended for years in both the laboratory and field. The nature of the compound used has depended on the type of information required, and this, in turn, has been determined by the objectives of the experiment.

The use of ^{14}C-labeled substrates is necessary to trace the fate of added organic matter in the various soil constituents. Furthermore, without the use of labeled substrates it is nearly impossible to distinguish the added C from soil C because the amount of C added is usually only 5–10% of the soil C. It is also important to emphasize that measurement of $^{14}CO_2$ evolution or residual organic ^{14}C accumulated in soil does not reflect the decomposition of the added substrate exclusively, except during early peak metabolic activity. During decomposition a significant portion of the substrate can be used for biosynthesis and these microbial products would also be subject to

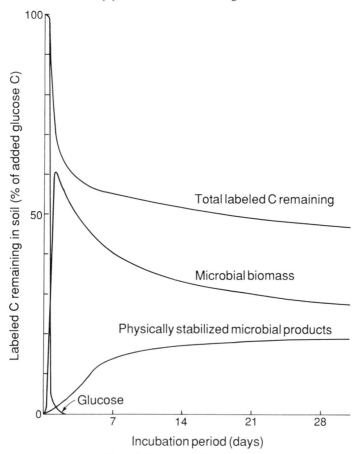

Fig. 1   Decomposition of glucose in a laboratory incubation.

degradation. Furthermore, the decomposition of added substrates may differ from that of the same compounds present in soil and generated *in situ* due to interactions of organic compounds with soil constituents.

^{14}C-glucose is often used as the substrate to label the microbial biomass and its metabolic products (Fig. 1). This approach has been used successfully to trace the fate of microbial products in fractions of soil organic matter, and in soil aggregates.

Labeled crop material is required if decomposition processes of agro-eco-systems or agronomic practices are to be examined. The movement of plant residue C has been traced into the microbial biomass during its exponential decline from the soil as CO_2 (Fig. 2). With time, the metabolism of actual plant residue C becomes insignificant and all that remains is resistant micro-

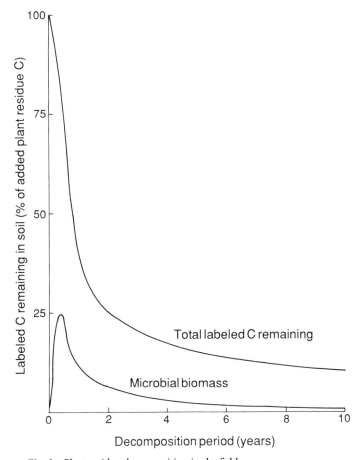

Fig. 2   Plant residue decomposition in the field.

bial and plant organic residues, and organic compounds that are protected from microbes by physical attachment to soil minerals (Table 1).

II. MATERIALS REQUIRED

A.   ¹⁴C-Labeled Organic Substances

In studying the decomposition of plant residues, the ^{14}C label in the plant residues should be distributed uniformly, or at least the specific activity of the label in the components of the plant tissue should be known. Erroneous conclusions may be drawn if the decay rates of ^{14}C-labeled components of

Table 1

Size and Dynamics of the Components of Soil Organic
Matter

	Relative size (% of total organic C)	$t_{1/2}$ (years)
Microbial biomass	1–3	1–2
Physically stabilized	30–50	50
Biologically recalcitrant	50–70	1500–3000

plant materials differ from those of unlabeled components, such as overestimating soil C transformations in the presence of the residue or in underestimating the extent of transformation of the residue. In terms of decomposability, plant tissue can be separated into three components: (1) labile cytoplasmic constituents and storage compounds such as starch; (2) medium-labile, cellulosic cell wall components; and (3) slowly degradable, lignified cell walls. These components differ in their rates of decomposition, and in the extent to which they are incorporated into microbial biomass and into the stabilized components of soil organic matter. Cytoplasmic constituents of plants are readily metabolized by soil microorganisms and a large proportion of the C is stabilized in recalcitrant microbial products. The aromatic compounds of plant lignins may undergo less modification and be incorporated directly into stable constituents of soil organic matter.

A common method used in producing labeled plant material is to grow plants in a controlled-atmosphere chamber maintained with a constant enrichment of $^{14}CO_2$ (see Warembourg and Kummerow, Chapter 2, this volume). However, the method requires a rather expensive growth cabinet and complicated regulating equipment for delivering CO_2. In addition, plant material prepared in this manner may have some drawbacks. The plants may not have been grown to maturity, and may not have the architecture and degree of lignification of field-grown material which develops under normal sunlight conditions, moisture stresses, and wind.

An alternative source of labeled plant residue can be acquired through the use of a repeated-pulse labeling technique. At regular intervals throughout the growing season plants are briefly exposed (0.5–2 hr) to $^{14}CO_2$. The quantity of $^{14}CO_2$ (Q: Bq ^{14}C) used to label the plant material depends on: (1) the desired final specific activity of the plant residues (SA: Bq ^{14}C/g C); (2) the efficiency of the labeling techniques for incorporation of labeled C into the plant tissue (E: 50–80%); (3) the loss of fixed labeled C through subsequent plant respiration (R: 30–60% of the incorporated labeled C); and (4) the total plant C production at harvest (Y_C: g C):

$$Q = \frac{SA}{E \times R} \times Y_C.$$

The fraction of the total $^{14}CO_2$ exposed at any one pulse should be proportional to the rate of plant matter formation at the time of the pulse. If ΔY_C is the increment of plant C (growth) between pulses, then the quantity (Δ: Bq) of $^{14}CO_2$ exposed at each pulse is:

$$\Delta = Q \times \frac{\Delta Y_C}{Y_C}.$$

The relationship of plant dry matter accumulation over time can be estimated from previous years' data, approximations from the literature, or by using a computer simulation model of plant growth. The organic C content of plant tissue is approximately 40% of its total dry matter.

The principal advantage of the pulse labeling procedure is that exposure to the stresses of the field environment produces plant residues typical of field-grown plants. An added benefit is that the decomposition of both shoot and root residues can be studied simultaneously. The limitation, however, is that the distribution of ^{14}C throughout the plant components is less uniform than that in plants grown under a continuous exposure to $^{14}CO_2$. Differences in specific activity occur due to the contribution of seedling C, variations in plant respiration, C translocation patterns, ability to label frequently enough during the period of exponential vegetative growth, inaccurate growth curve prediction, and other factors. The uniformity of ^{14}C should be confirmed by a plant tissue analysis such as described by Warembourg and Kummerow (Chapter 2, this volume) or by neutral and acid detergent fiber extraction (Van Soest and Robertson, 1979).

The maximum specific activity of the labeled plant material produced is limited due to irregular plant growth at high levels of exposure to radiation. Exposure to CO_2 with specific activities about 1 MBq ^{14}C/g C have been observed to allow normal growth in a number of annual plant species, although seed produced may suffer faster than usual decline in viability with storage. Although growth can be reduced, maize (*Zea mays* L.) and soybeans (*Glycine max* L.) can withstand up to 37 MBq ^{14}C/g C. Since microorganisms can withstand higher levels of radioactivity, tracing the fate of ^{14}C into fractions of soil organic matter is limited by the upper level of the specific activity attainable in plants.

B. Laboratory Soil Incubation Studies

With the addition of ^{14}C-labeled substrates to the soil, information on the amount and rate of mineralization of the substrate can be obtained. The rate of decomposition of organic substrates in soil is influenced by temperature, O_2 supply, water potential, pH, inorganic nutrients, and the C:N ratio of the added substrate, all of which can be controlled in laboratory incubations. Measurements of soil respiration have been widely used to assess the general activity of the soil biomass.

The amount of ^{14}C-labeled substrate added to the soil should be precisely known. The specific activity of the nutrient solution should be rechecked just prior to addition to the soil. In addition, the ^{14}C-labeled substrate remaining in the container in which the soil and nutrient solution were mixed should be measured.

In order to maintain an active, viable biomass, the water content of the soil must be considered. The incubation conditions in the laboratory may range from dry to saturated depending on the particular experiment. Rewetting previously air-dried soil can have a substantial effect on mineralization of soil organic matter. In addition, alkaline soils, and some soils that have been air dried and rewetted, can release substantial amounts of CO_2 from abiotic as well as biological reactions. Incubation of soil at (or near) the desired soil water potential of the experiment for $1-2$ weeks prior to substrate addition avoids this complication.

Microbial metabolism involving the aerobic oxidation of organic C to CO_2 occurs in the range of soil water potentials encountered for growth of higher plants, namely, from -33 to -1500 kPa. We have found -45 to -60 kPa to be optimal for soils with less than 40% clay. In soils with high clay contents it may be difficult to thoroughly mix a substrate into a soil at high water potentials (between 0 and -50 kPa) because of clumping of soil particles; moistening the soil with finely crushed ice reduces this problem. If an aqueous labeled solution is to be used, it is usually easier to add it to the soil first and then adjust the soil water potential to the desired value. The moistened soil should be weighed regularly throughout the experiment and water added to maintain a constant soil water potential.

Adequate aeration of the incubating soil is an important consideration for maximizing oxidation rates of organic substrates. The soil may be placed as a shallow layer in the incubation vessel. Diffusion of oxygen to the site of microbial activity will be impeded if this layer is too thick. Cylinders constructed out of wire mesh have been used in our incubation studies to ensure that aerobic conditions are maintained and the distance that oxygen has to diffuse is minimal. These are made with double layers of 1-mm (No. 18) wire mesh (5 cm diameter by 15 cm height). At regular intervals during the incubation, when not more than 20% of the O_2 has been used up, the sealed incubation vessels should be opened to replenish O_2 levels. Adjustments for loss of soil water can also be made at these times.

C. Microplots for Field Studies

Traditionally, microplot techniques have been used in field studies to provide field incubation conditions while limiting lateral movement and dilution of labeled material with the surrounding soil. Microplots may consist of 15–30-cm diameter cylinders driven 30–40 cm into the soil, or of plots 0.5 m² to several square meters in area, surrounded by some form of buried barrier. The size of the microplot determines the exposed surface area; this is an important consideration for representation of field conditions, simulation of soil management, or for providing sufficient rooting volume if plants are grown. Cylinders 15–30 cm in diameter driven 30–100 cm into the soil may suffice for studying the movement of ^{14}C into components of soil organic matter. If after a sufficient period of incubation the entire cylinder of soil is removed for analysis, four replicates provide a sufficient number of observations to describe the dynamics of ^{14}C microbial biomass and loss of $^{14}CO_2$. Cylinders provide an additional advantage in that highly labeled substrates can be used without prohibitive expense.

Studies of processes occurring under large plants or row crops may require larger microplots. For example, maize planted in 75-cm row widths requires a surface area of roughly 75 cm $\times$ 20 cm per plant. If the effect of the crop row position is important, such as in the distribution and decomposition of root residues, then the microplots may have to be large enough to include more than one row with two to three plants per row. Replication of observations (rows) by using a transect microplot across several rows maximizes the number of observations per unit area of soil treated with labeled material.

When large microplots are used, either a systematic or random soil core sampling procedure may be necessary rather than removing all of the microplot soil. Systematic sampling should be used when a systematic variation across the microplot is suspected. Row crops might have soil samples taken in the row and at interrow positions. Random sampling may be used if variability is suspected across a microplot, but in no regular pattern. This can occur when residues are placed on the soil surface.

If the whole microplot is not sampled at one time, coarse grinding or chopping of the material may be necessary to provide a more even distribution of the plant residues and allow representative sampling without excessive soil removal. A representative sample may be obtained from bulking many randomly taken small subsamples. The same microplot may be repeatedly sampled over time providing only 1–2% of the soil volume is removed and the sampling holes are refilled with plastic pipe or sand.

A nylon mesh can be used to keep surface residues in place. Water action, confounded by variations in microtopography of the soil surface, and wind action can redistribute or remove surface residues from the microplots.

III. ANALYTICAL PROCEDURES

A. Microbial Biomass

1. Introduction

Measurements of the flow of C through the microbial biomass in soil amended with ^{14}C-labeled plant residues can be measured using the chloroform ($CHCl_3$) fumigation-incubation method (Jenkinson, 1966). Fumigation of soil samples with $CHCl_3$ vapor causes a flush of decomposition during a subsequent 10-day incubation. The flush of CO_2, due to mineralization of microbial cells killed by the fumigation treatment, is directly proportional to the size of the microbial biomass C pool. Alternatively, microbial biomass C released by the fumigation treatment can be extracted directly with a K_2SO_4 solution and measured as organic C (Ladd and Amato, 1988; Vance et al., 1987). The ^{14}C content of the microbial biomass can be measured using either method; however, the $CHCl_3$ fumigation-extraction (F.D.E.) method will be described in detail because it is relatively easy and much more rapid.

2. Materials

a. Fumigation

- A fumehood.
- A large desiccator. The desiccator should be inert to $CHCl_3$ vapor, be of a dry-seal type, and be able to withstand a high vacuum without implosion; thick-walled glass construction is suitable.
- Paper towels.
- 25 ml of freshly purified $CHCl_3$ (Note: Reagent-grade $CHCl_3$ generally contains a stabilizing agent, alcohol, which must be removed prior to use. We routinely use $CHCl_3$ stabilized with heptachlor epoxide at 10 pg/ml, and obtain data similar to that with purified $CHCl_3$.)
- Boiling chips.
- 1 weighing container and 2 100-ml glass bottles with lids per assay.

b. Extraction

- 2 Whatman #5 filter papers per assay.
- 2 × 50 ml 0.5M K$_2$SO$_4$ per assay.
- 2 50-ml plastic vials per assay for filtrate.

c. Scintillation Counting

- 2 20-ml scintillation vials per assay.
- 10–15 ml scintillation cocktail.

3. Procedure

a. Preparation of Soil Samples

- Pass a soil sample through a 2-mm mesh sieve and mix thoroughly.
- Weigh 3 portions of the soil, 15–25.0 g each, into the weighing containers, and into 2 100-ml glass jars, one jar to be fumigated, and one control jar to be extracted immediately.
- Dry the soil in the weighing containers (with the lid off) in an oven at 105°C for at least 24 hr or until a constant oven-dry weight is achieved. Cool in a desiccator, reweigh, and determine the water content of the soil sample.

b. Fumigation of Soil Samples (Note: Procedures releasing CHCl$_3$ fumes should be conducted in a fumehood.)

- For the fumigation treatment, place the glass sample jars containing the soil in a desiccator lined with moistened paper towels together with a beaker of CHCl$_3$ and a few boiling chips. A large desiccator may contain about 25 samples simultaneously.
- Evacuate the desiccator in a fumehood until the CHCl$_3$ boils, and continue evacuating for 2 min.
- Place the desiccator in the dark for 18–24 hr.
- Open the desiccator and remove the beaker of CHCl$_3$ and the paper towels.
- Remove all residual CHCl$_3$ vapor from the soil samples by repeated evacuation, usually 3–6 times, using first an aspirator pump followed by a two-stage rotary oil pump capable of drawing a vacuum of 10^{-5} kPa, for a total of 15–20 min of evacuation.

c. Extraction of Microbial Biomass C

- To the jars containing the control and fumigated subsamples, add 50 ml of 0.5M K$_2$SO$_4$, i.e., a soil:extractant ratio of 1:2.

- Cap the jars and place on an oscillating shaker for 1 hr.
- After shaking, filter the soil suspension and collect the filtrate. Avoid excessive evaporation. Cap and store the filtrate at 4°C for not more than 2–3 days, otherwise freeze until ready for analysis.

d. Determination of Microbial Biomass ^{14}C

- Pipet an aliquot (e.g., 2 ml) of the K_2SO_4 extract into a liquid scintillation vial containing an aqueous compatible cocktail.
- Count in a liquid scintillation counter to determine ^{14}C activity as Bq/aliquot volume of soil extract.

e. Total Microbial Biomass C

- Organic C in the K_2SO_4 extracts can be determined using automated equipment (e.g., Method No. 455-76W/A, Technicon Industrial Systems, Tarrytown, New York) or by dichromate digestion (Jenkinson and Powlson, 1976a).
- For digestion by dichromate, an 8-ml aliquot of the K_2SO_4 extract is added to a mixture of 0.2M $K_2Cr_2O_7$ (2 ml), 18M H_2SO_4 (10 ml), 14.7M H_3PO_4 (5 ml) and HgO (70 mg) and boiled under refluxing conditions for 30 min. The excess dichromate remaining is determined by titration with 0.017M ferrous ammonium sulphate using ferroin as an indicator.

4. Calculation

a. Soil Water Content

$$W_s(\%) = \frac{\text{soil wet weight} - \text{soil oven-dry weight}}{\text{soil oven-dry weight}} \times 100$$

b. Weight of Soil Sample (Oven-dry Weight Equivalent) Taken for Microbial Biomass Measurements (M_s)

$$M_s = \frac{\text{soil wet weight} \times 100}{(100 + W_s(\%))}$$

c. Total Volume of Solution in the Extracted Soil (V_s)

$$V_s = \text{soil wet weight} - \text{soil oven-dry weight} + \text{extractant volume}$$

d. Weight of Organic C Derived from the Added ^{14}C Substrate in an Aliquot of the Fumigated or Unfumigated Soil Sample Extract

$$\frac{\text{substrate-derived C weight}}{\text{aliquot volume of soil extract}} = \frac{\text{Bq/aliquot volume of soil extract}}{\text{SA}}$$

where

$$SA = \text{substrate specific activity}$$

$$= \frac{Bq}{\text{substrate C weight}}$$

e. Total Weight of Organic C or Organic C Derived from the Added ^{14}C Substrate in the Fumigated or Unfumigated Soil Samples

$$OC_F, OC_{UF} = \frac{\text{organic C weight}}{\text{aliquot volume}} \times \frac{V_S}{M_S}$$

$$= \frac{\text{organic C weight}}{\text{oven-dry weight soil}}$$

f. Microbial Biomass in the Soil

microbial biomass C, total or derived from added ^{14}C substrate
$$= (OC_F - OC_{UF})/k_C$$

where $k_C = 0.35$ and represents the efficiency of extraction of microbial biomass C.

5. Comments

This method is relatively precise; 4 replicate determinations on the same soil sample should be able to detect differences in microbial biomass of 5–10% at a 0.05% level of probability. Across different soil types, the method may be less accurate due to variations in $k_C = 0.35 \pm 0.05$.

Chemiluminescence effects can be avoided when counting ^{14}C in K_2SO_4 extracts (Step 3d) by storage of the samples in the dark for 7–14 days prior to counting to let background counts subside. Background counts will vary between soil samples, but are assumed to be similar for fumigated and unfumigated subsamples drawn from a well-homogenized soil sample. Background counts do not enter into the calculation of microbial biomass ^{14}C.

Microbial biomass assays can be made still more rapid by direct addition of 1 ml of $CHCl_3$ to suspensions of soil in the 50 ml of K_2SO_4 solution and then shaking for 0.5 hr. After filtration, $CHCl_3$ is expelled by bubbling CO_2-free air through the filtrate for 30–45 sec. This method is as accurate as the F.D.E. method but less microbial biomass C is recovered ($k_C = 0.20$); this may be an important factor if microbial biomass ^{14}C levels are low.

Higher levels of recovery of biomass C ($k_C = 0.45$) can be obtained by using the original fumigation-incubation method and measuring $^{14}CO_2$ trapped in a solution of NaOH. However, this method has proven to be inappropriate for soils with a pH <4.2 or for those containing significant levels of unmetabolized labile organic substrates. For soils with a low pH, microbial C can be measured successfully using the F.D.E. method (Vance *et al.*, 1987).

B. Total Soil Organic ^{14}C

1. Introduction

Two common methods, dry and wet oxidation, are available for ^{14}C determination in soils. For the dry oxidation method, commercial combustion furnaces are available. The LECO induction furnace (LECO Corporation, St. Joseph, Michigan) has been thoroughly tested for completeness of soil organic matter combustion, and Cheng and Farrow (1976) have described minor modifications to a LECO furnace to measure ^{14}C down to approximately 300 Bq ^{14}C/g soil. Effluent gas from the furnace is passed through a purifying train before being scrubbed for ^{14}CO$_2$ in an NaOH solution.

Methodology for the wet oxidation method will be described because the equipment is less costly, and the detection limit is approximately 15 Bq ^{14}C/g soil. The procedure outlined is essentially that of Amato (1983). Soil or plant material is weighed into a digestion tube, and a digestion mixture is added along with a vial of CO$_2$-trapping agent supported above the digestion mixture on a glass rod pedicel (Fig. 3). The tube is capped, heated, cooled overnight, and absorbed ^{14}CO$_2$ is counted in the trapping solution.

2. Materials

a. Digestion Apparatus

- A block digester.
- Digestion tubes (250 × 25 mm), with thick walls (1.5 mm) to withstand pressure.
- A subaseal (#49) to stopper the digestion tube.
- A 2-ml glass vial, or a 10-ml graduated tube (if total C is desired).
- A glass support rod (5-mm diameter × 170-mm length) bent at one end to support a glass vial, or a shorter rod (95 mm) if graduated tubes are used.

b. Reagents

- 1 ml ethanolamine, or 5 ml 2M NaOH to trap CO$_2$.
- Digestion mixture:
      189.4 g CrO$_3$
      250 ml 14.7M H$_3$PO$_4$
      500 ml 18M H$_2$SO$_4$

c. Scintillation Cocktail for Counting Ethanolamine (1 l)

- 700 ml toluene
- 400 ml methoxy ethanol

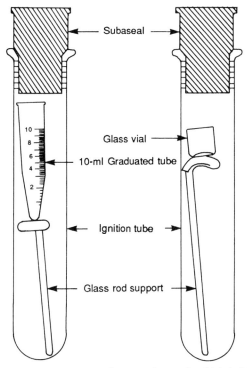

Fig. 3 Digestion tube set-up for total and labeled C determination.

- 4 g 2,5-diphenyloxazole (PPO)
- 100 mg 1,4-bis(5-phenyloxazole-2-yl) benzene (POPOP)

3. Procedure

a. Preparation of the Digestion Mixture

- Place the H_3PO_4 and H_2SO_4 into a 500-ml Pyrex Erlenmyer flask and heat to 145–150°C, using a hot plate and a magnetic stirrer to facilitate dissolution of the CrO_3. When 150°C has been reached, turn off the heat. As the temperature starts to fall from 150°C, add CrO_3 and continue to stir during cooling. Keep covered with an inverted beaker to prevent water vapor from entering the flask since water diminishes the oxidizing efficiency of the mixture. The digestion mixture should be brown colored; prolonged heating produces a green solution of oxidized chromium.

b. Digestion of Plant and Soil for Total ¹⁴C

- Weigh into the digestion tube 0.3–1 g of soil (oven dried at 60°C and ground to finer than 0.15 mm), or 0.04 g of plant material (oven dried at

60°C and ground to finer than 0.25 mm). Samples should contain less than 20 mg C.

- Prepare the CO_2 trap by adding 1 ml of ethanolamine to the 2-ml vial.
- Add 6 ml of cool digestion mixture to each tube, quickly add the glass support rod and CO_2 trap, and stopper the tube with a subseal.
- Place the tube in a heating block for 1 hr at 130°C (but no higher), then cool and leave overnight to allow complete CO_2 absorption. **CAUTION: The contents of the tube are under pressure.**
- In a fumehood and wearing protective eyeglasses, remove the CO_2 trapping vial from the digestion tube and place it into a glass scintillation vial. Add 11 ml of scintillation cocktail, cap tightly with a foil-backed plastic cap and swirl gently until the contents are thoroughly mixed; the cocktail should appear clear. Storage of the samples for 24 hr prior to counting slightly increases the counting efficiency.

c. Digestion of Plant and Soil for Total ^{14}C and Total C

- Weigh plant or soil and add to a digestion tube as in 3b.
- Prepare the CO_2 trap by adding 5 ml of 2M NaOH to a 10-ml graduated tube.
- Add the digestion mixture, CO_2 trap, and stopper as in 3b.
- Digest, cool, and store as in 3b.
- Remove the CO_2 trap from the digestion tube and add an aliquot (e.g., 0.1 ml) of the NaOH solution to a scintillation cocktail.

d. Measurement of Total C

- After removing the aliquot for scintillation counting in c. above, dilute the remaining NaOH solution in the graduated tube to 10 ml with CO_2-free distilled water and seal the tube to prevent further exposure to CO_2.
- Total C can be determined by titration of the NaOH solution using the method described by Anderson (1982).

4. Comments

This method of wet combustion gives results comparable to those using dry combustion techniques, and greater than 98% of the organic C in standards can be recovered. A correction for samples processed with ethanolamine has to be made as it is 2.5% less efficient at absorbing $^{14}CO_2$ compared to those using NaOH. It is also recommended that for a given amount of unknown C and ^{14}C content, duplicate subsamples should be weighed in two known amounts, e.g., x and 0.5x, and that the measured radioactivity should be confirmed to be directly proportional to the amounts contained. This circumvents problems that may arise with soils of high organic C content exceeding the absorbing capacity of the ethanolamine absorbent.

Each batch of samples should include a ^{14}C standard to determine the efficiency of the digestion mixture, and a digestion tube containing no sample for measurements of background counts in the scintillation cocktail. For analysis of the ^{14}C content of a set of soil samples, a soil of known ^{14}C content is a reliable standard; compared to soil, standards of methyl methacrylate are less easily oxidized, whereas lactic acid standards are more easily oxidized. The ^{14}C of the standard soil can be calibrated by comparison to the oxidation of ^{14}C-methyl methacrylate standards, providing an 80% counting efficiency is obtained after digestion of a ^{14}C-methyl methacrylate standard and capturing ^{14}CO$_2$ in ethanolamine. A ^{14}C counting efficiency of 80% indicates that the oxidation of the ^{14}C-methyl methacrylate standard is complete. The standard soil (~ 100 g) should be oven dried (105°C, 24 hr) and homogenized by grinding to pass through a 0.15-mm sieve. If a standardized soil is used, there should be at least two replicates per batch of samples.

Absorption of ^{14}CO$_2$ in ethanolamine is particularly useful for samples of low radioactivity. In cases where the activity is low and total organic C is also desired, total organic C can be measured on a separate subsample. Premixed cocktails obtained from commercial suppliers for counting ^{14}CO$_2$ absorbed in NaOH usually have counting efficiencies of about 85%.

To determine the organic C in calcareous soils, total C measurements must correct for the inorganic C present. Inorganic C can be determined by adding 9 ml of $0.33M$ FeSO$_4$ to the soil in a digestion tube, and freezing the suspension. While still frozen, 1 ml of $9M$ H$_2$SO$_4$ is added, an NaOH trap inserted, and the tube sealed with a stopper. The sample is then thawed and the tube is placed in a digesting block at 50°C for 1 hr. The sample is removed to cool and left overnight to allow complete absorption of the acid-released CO$_2$, prior to titration of the NaOH solution for the determination of inorganic C.

C. Rates of Evolution of ^{14}C-Labeled CO$_2$

The rate of evolution of ^{14}C-labeled CO$_2$ can be measured in open or closed systems. In a typical open incubation, the soil sample is placed in an aeration train with a trap for collecting the evolved CO$_2$. Several types of incubation units have been described, some of which have elaborate and sophisticated automatic devices for controlling temperature and humidity and for continuously monitoring O$_2$ and CO$_2$ concentrations. Closed systems are generally more simple but require regular opening to the atmosphere to ensure that O$_2$ does not become limiting. A convenient and inexpensive procedure for estimating the rate of ^{14}CO$_2$ evolution is to incubate the soil in an airtight preserving jar (1–2 l capacity) and to trap the $^{14}CO_2$ in an alkali solution, such as NaOH. An aliquot, usually 1–2 ml, of the $NaOH/Na_2{}^{14}CO_3$ solution is taken for ^{14}C determination by liquid scintillation counting. Total CO_2 is measured by precipitating $BaCO_3$ from the reacted alkali with $BaCl_2$, and titrating unreacted alkali with standard acid. An important consideration is the strength of the alkali solution: it must capture all of the CO_2 evolved at peak respiration rates and yet be of a concentration sufficiently low as not to exceed the capacity of the scintillation cocktail to emulsify the solution. Chemiluminescence from the scintillation cocktail, caused by alkali solutions, can be minimized by storage of the sample counting vials in the dark for 24–48 hr.

IV. AMOUNT OF ^{14}C REQUIRED

A. Introduction

Calculations of the minimum requirement of ^{14}C substrate or of specific activity (SA: Bq/substrate C weight) requires an understanding of the processes under study in a particular experiment, the laboratory procedures used for analysis, and the minimum quantity of label that can be measured in the soil or some soil organic fraction of interest. For some procedures of fractionation of soil organic matter and of microbial products, the minimum level of ^{14}C may have to be determined empirically by a pilot experiment. Factors to consider in deciding on the appropriate substrate SA include: (1) the rate of substrate addition to the soil (R: g C/g soil); (2) the proportions of substrate C that are transformed into different soil organic matter components, i.e., the partitioning coefficient (PC: no units) for each fraction of interest; (3) the analytical efficiency (AE: g soil/cocktail) of the laboratory procedure, which indicates whether the procedure has concentrated or has diluted an organic fraction C; and (4) the minimal activity (MA: Bq/cocktail) desired in a scintillation cocktail for efficient counting. Thus,

$$SA = MA/(AE \times PC \times R) \text{ or}$$

$$SA = MA/C$$

where C is the quantity of fraction C in the cocktail.

An example of the calculations necessary to determine the substrate SA is shown later together with the usual calculations for detection of labeled C in microbial biomass, in particle size fractions of soil, and in residual stabilized soil organic matter.

B.   Rate of Substrate Addition to Soil

The amount of substrate C added to a soil is usually dictated by the ecosystem under study and related to the annual litterfall. Using an example from agriculture, a maize crop could produce up to 6000 kg/ha of dry matter in stover. If the C content of the plant residue is 40%, this represents an addition of 2400 kg C/ha to the soil. Tillage would incorporate this crop residue into the soil to a depth of about 0.15 m. If a soil bulk density of 1.3 Mg/m³ is assumed, 1 ha of soil 0.15 m deep has a mass of 1950 Mg and the rate of crop residue addition is 1.2 kg C/Mg soil.

In experiments designed to study the microbial biomass, 1 mg substrate C/g soil is commonly added. If glucose is used as the substrate, about 650 μg labeled microbial biomass/g soil is produced. This typically is of the same order of magnitude as the size of the native (unlabeled) microbial biomass in cultivated soils, without changing drastically the nature of the microbial populations or their decay rates. However, even with the addition of a simple substrate such as glucose, not all of the microbial biomass becomes uniformly labeled. This is because only a portion of the native biomass, an "active biomass," responds and grows on the added substrate.

Microorganisms use much smaller amounts of added substrate, 200–500 μg C/g soil, for maintenance rather than for growth, resulting in a minimal increase in biomass. Larger additions of substrate can confound the subsequent dynamics of labeled C due to wide fluctuations in microbial populations and to severe limitations in the availability of nutrients other than C which are required for microbial growth.

C.   Estimation of the Fate of Substrate C and the Partitioning Coefficient

During decomposition, substrate is assimilated by the microbial biomass, and eventually most will be mineralized and evolved from the soil as CO_2 (Fig. 1 and 2). The residual C derived from the substrate declines exponentially to levels accounted for by C compounds either protected from mineralization by their innate chemical recalcitrance or by their association with soil minerals. This residual C may represent 5–20% of the initial substrate C added and it is relatively stable.

The residual C recovered in the microbial biomass increases to a maximum, about 65% of input C if glucose is the substrate (Fig. 1) and up to 25% if mature plant residue is the substrate (Fig. 2). Microbial biomass C then declines exponentially.

Redistribution of substrate and residues may occur among organic matter fractions, particle size fractions, and aggregate size fractions. Studies have shown that when either simple labeled compounds or complex substrates such as plant material are added to soil and allowed to incubate, then label is found in all of the classical fractions (humic acid, fulvic acid, and humin) of soil organic matter obtained by alkali extraction. Water-soluble labeled C can be relatively high, from 15 to 30% of the total labeled C recovered, in soils incubated for short periods (<90 days) decreasing to <10% after 5–6 years of incubation.

Clay fractions usually contain more than 50% of the residual labeled C. The proportions of the labeled C in silt-size fractions increase with time due to (1) a greater stability of silt-bound organic matter, or (2) a transfer of organic matter from clay to silt-size particles.

Substrate C enters different components of the soil to varying degrees and the proportion of substrate C found in a given fraction can be called the partitioning coefficient (PC). The researcher has to estimate the least proportion of substrate C to be detected in a given fraction. For example, in the calculations below, the limit is set to detect residual C at a minimum PC = 0.15, microbial biomass C at a minimum PC = 0.10, and particle- or aggregate-size fraction C at a minimum PC = 0.05.

D. Calculation of Analytical Efficiency

Analytical efficiency (AE) is the effective mass of soil measured in each assay of the labeled organic fraction. It is a function of the laboratory procedures used to isolate that organic fraction and the size of the aliquot of the organic fraction C mixed into the scintillation cocktail. For example, it is affected by the mass of soil that can be analyzed, how effectively the organic fraction is isolated from the soil, and whether the organic fraction is diluted or concentrated by the carrying medium mixed into the scintillation cocktail. A specific AE must be carefully derived for each analytical procedure. Some example calculations follow.

1. Total Residual C Remaining in Soil

Assume that 1 g of soil is digested, that the combustion of organic matter is complete, and that the evolved $^{14}CO_2$ is trapped in 1 ml ethanolamine, which is transferred *in toto* to a scintillation cocktail. Thus, the critical aspects of the procedure are:

1 g of soil is analyzed per digestion, or 1 g soil/ml ethanolamine
since all of the ethanolamine is added to the scintillation cocktail, then:
1 ml ethanolamine/cocktail.

From the above, AE is:

AE = 1 g soil/ml ethanolamine $\times$ 1 ml ethanolamine/cocktail
    = 1 g soil/cocktail.

Therefore, the residual substrate C from 1 g soil is measured in each scintillation cocktail.

2. Residual C in Particle-Size Fractions

Assume that sand, silt, and clay particle-size fractions have been isolated following treatment with ultrasonic energy and that the total organic C in each size fraction as well as the residual labeled C is measured. A wet combustion method is used to oxidize 1 g of each size fraction and the evolved CO_2 is absorbed in 5 ml of 2M NaOH. A 0.1-ml aliquot of the NaOH solution is counted in a scintillation cocktail. The remaining NaOH is titrated to determine total C. Digestion of the organic C in each soil particle size fraction is assumed to be complete. Critical aspects of the procedure are:

1 g of soil is analyzed per digestion, or 1 g soil/ 5 ml NaOH solution
NaOH solution is subsampled for the scintillation cocktail:
0.1 ml NaOH/cocktail.

From the above, AE is:

AE = 1 g soil/5 ml NaOH $\times$ 0.1 ml NaOH/cocktail
    = 0.02 g soil/cocktail.

The residual substrate C from a 0.02-g soil aggregate fraction is analyzed in each scintillation cocktail.

3. Residual C in Microbial Biomass

Assume that microbial biomass is determined on a soil sample using the F.D.E. method: 20 g soil wet weight (15 g soil, oven-dry weight) is fumigated, then extracted with 30 ml of 0.5M K_2SO_4. A 2-ml aliquot of the extract is added to the scintillation cocktail. The critical aspects of the procedure are:

Dilution of microbial C by the extracting solution:
The total volume of solution in the extracted soil sample (V_S) is
   V_S = soil wet weight − soil oven-dry weight + extractant volume
       = 20 − 15 + 30 (assuming the density of water is equal to 1 g/ml)
       = 35 ml.
Therefore:
   15 g soil/35 ml extracting solution

Aliquot volume of extract added to the scintillation cocktail, e.g., 2-ml aliquot of extract cocktail.

From the above, AE is:

$$AE = \frac{2 \text{ ml aliquot of soil extract}}{\text{cocktail}} \times \frac{15 \text{ g soil}}{35 \text{ ml extract}}$$

$$= \frac{0.86 \text{ g soil}}{\text{cocktail}}.$$

Thus, in this example of measuring biomass ^{14}C, 0.86 g soil is analyzed in each vial of scintillation cocktail. It is important to remember that in this procedure the extracted C is only 35% of the total soil microbial C, that is $k_C = 0.35$. The total microbial biomass C in the soil is determined using the calculations presented in Section III,A,4.

E. Bq of ^{14}C in the Scintillation Cocktail

For a relatively short counting time of 10 min, 1000 counts per minute (CPM) are required in order to have a confidence level of 95% for detecting a difference of 20 CPM.

The counting efficiency (CE) of ^{14}C must be determined for each analytical procedure and scintillation cocktail used. In the example calculations below, the assumption is that the counting efficiency is 85%, which is used to convert the sample CPM to disintegrations per minute (DPM):

$$\text{sample DPM} = \frac{\text{sample CPM}}{\text{CE}}$$

$$= \frac{1000 \text{ CPM}}{0.85}$$

$$= 1176 \text{ DPM}$$

Therefore, the minimum activity (MA) of ^{14}C, expressed in Bq, in the cocktail should be:

$$MA = \frac{1176 \text{ DPM}}{60 \text{ DPM/Bq}}$$

$$= 20 \text{ Bq}$$

(Note: A Becquerel is activity equal to one disintegration per second.)

F. Minimal ^{14}C Requirements

1. Total Residual C in Soil

The amount of labeled C in the cocktail (C) is:

$$C = AE \times PC \times R$$

$$= \frac{1 \text{ g soil}}{\text{cocktail}} \times \frac{0.15 \text{ mg } {}^{14}C \text{ residue C}}{\text{mg } {}^{14}C \text{ substrate C}} \times \frac{1 \text{ mg substrate C}}{\text{g soil}}$$

$$= \frac{0.15 \text{ mg } {}^{14}C \text{ residue C}}{\text{cocktail}}.$$

The specific activity of the substrate C is:

$$SA = MA/C$$

$$= \frac{20 \text{ Bq/cocktail}}{0.15 \text{ mg } {}^{14}C \text{ residue C/cocktail}}$$

$$= \frac{133 \text{ Bq}}{\text{mg } {}^{14}C \text{ residue C}}.$$

Therefore, the minimum SA of the added substrate must be 133 Bq/mg C to be able detect at least 15% of the labeled residual C in the soil.

2. Residual C in Particle-Size Fractions

Substrate C in the cocktail (C) is:

$$C = AE \times PC \times R$$

$$= \frac{0.02 \text{ g soil}}{\text{cocktail}} \times \frac{0.05 \text{ mg residue C}}{\text{mg substrate C}} \times \frac{1 \text{ mg substrate C}}{\text{g soil}}$$

$$= \frac{1 \times 10^{-3} \text{ mg residue C}}{\text{cocktail}}.$$

Specific ^{14}C activity of the substrate C is:

$$SA = MA/C$$

$$= \frac{20 \text{ Bq/cocktail}}{1 \times 10^{-3} \text{ mg residue C/cocktail}}$$

$$= \frac{20,000 \text{ Bq}}{\text{mg residue C}}.$$

Therefore, the minimum SA of the added substrate must be 20,000 Bq/mg C to be able to detect at least 5% of the labeled residual C in a particular particle size fraction. This SA would typically allow study of the labeled C in the sand-size fractions as they have a PC = 0.05; for the clay fraction PC would be much higher, ~0.50.

3. Residual C in Microbial Biomass

Substrate C in the cocktail (C) is:

$$C = AE \times PC \times R$$

$$= \frac{0.35 \text{ g soil}}{\text{cocktail}} \times \frac{0.1 \text{ mg biomass C}}{\text{mg substrate C}} \times \frac{1 \text{ mg substrate C}}{\text{g soil}}$$

$$= \frac{0.035 \text{ mg residue C}}{\text{cocktail}}.$$

The specific activity of the substrate C is:

$$SA = MA/C$$

$$= \frac{20 \text{ Bq/cocktail}}{0.035 \text{ mg residue C/cocktail}}$$

$$= \frac{571 \text{ Bq}}{\text{mg residue C}}$$

Therefore, the minimum SA of the added substrate must be 571 Bq/mg C to be able to detect at least 10% of the labeled residual C in the soil microbial biomass.

V. USE OF ¹⁴C TO ESTIMATE POOL SIZES

While ^{14}C tracers have been widely used to study the processes involved in organic C dynamics in microbe/plant/soil systems, there has been little research describing the use of ^{14}C to measure the sizes of the components of soil organic matter. The main reason for this has been a lack of methods for isolating fractions of soil organic matter that are meaningful in terms of the biology of the system. Furthermore, organic matter stabilization involves both physical and chemical mechanisms, necessitating the application of both physical and chemical techniques to separate organic matter components.

Recently developed techniques for measurement of the soil microbial biomass have made it possible to isolate the pool of organic C in the biomass. By ^{14}C-labeling the microbial biomass using a readily decomposable substrate such as glucose, it is possible to estimate the size *in situ* of the total (labeled + unlabeled) soil microbial biomass (Voroney and Paul, 1984).

Estimating a particular pool size is done by labeling the pool of interest and measuring its specific activity. On the conditions that: (1) the quantity of label in the pool is known, and (2) the specific activity of the pool can be measured, the total unlabeled (^{12}C) pool size can be calculated using isotopic dilution techniques:

$$\text{total } ^{12}C \text{ in pool} = \frac{\text{total } ^{14}C \text{ in pool}}{\text{specific activity of pool } (^{14}C/^{12}C)}$$

This was the approach used by Jansson (1958) to calculate the sizes of the active and passive organic matter pools and, more recently, it has been used by Bottner (1985) to study the active and dormant fractions of the microbial biomass. The sizes of other components of soil organic matter could be similarly determined.

REFERENCES

Amato, M. (1983). Determination of carbon ^{12}C and ^{14}C in plant and soil. *Soil Biol. Biochem.* **15**, 611–612.

Anderson, J. P. E. (1982). Soil Respiration. In "Methods of Soil Analysis, Part 2. Chemical and Microbiological Properties" (A. L. Page, Ed.), pp. 831–871. A.S.A. S.S.S.A., Madison, WI.

Bottner, P. (1985). Response of microbial biomass to alternative moist and dry conditions in a soil incubated with ^{14}C- and ^{15}N-labelled plant material. *Soil Biol. Biochem.* **17**, 329–337.

Cheng, H. H., and Farrow, F. O. (1976). Determination of ^{14}C-labelled pesticides in soils by a dry combustion technique. *Soil Sci. Soc. Am. J.* **40**, 148–150.

Jansson, S. L. (1958). Tracer studies on nitrogen transformations in soils with special attention to mineralization-immobilization relationships. *Ann. Roy. Agric. Coll., Sweden* **24**, 101–361.

Jenkinson, D. S. (1966). Studies on the decomposition of plant material in soil. II. Partial sterilization of soil and the soil biomass. *J. Soil Sci.* **17**, 280–302.

Jenkinson, D. S. (1977). Studies on the decomposition of plant material in soil. IV. The effect of rate of addition. *J. Soil Sci.* **28**, 417–423.

Jenkinson, D. S., and Powlson, D. S. (1976a). The effects of biocidal treatments on metabolism in soil—I. Fumigation with chloroform. *Soil Biol. Biochem.* **8**, 167–177.

Jenkinson, D. S., and Powlson, D. S. (1976b). The effects of biocidal treatments on metabolism in soil—V. A method for measuring soil microbial biomass. *Soil Biol. Biochem.* **8**, 209–213.

Ladd, J. N., and Amato, M. (1988). Relationships between biomass ^{14}C and soluble organic ^{14}C of a range of fumigated soils. *Soil Biol. Biochem.* **20**, 115–116.

Vance, E. D., Brookes, P. C., and Jenkinson, D. S. (1987). An extraction method for measuring soil microbial biomass C. *Soil. Biol. Biochem.* **19**, 703–708.

Van Soest, P. J., and Robertson, J. B. (1980). Systems of analysis for evaluating fibrous feeds. In "Standardization of Analytical Methodology for Feeds" (W. J. Pigden, C. C. Balch, and M. Graham, eds.), pp. 49–60. International Development Research Centre & International Union of Nutritional Sciences. Workshop Proceedings, Ottawa, Canada. March 12–14, 1979.

Voroney, R. P., and Paul, E. A. (1984). Determination of k_C and k_N *in situ* for calibration of the chloroform fumigation-incubation method. *Soil Biol. Biochem.* **16**, 9–14.

6

Environmental Toxicology: Degradation of Herbicides

Frederick T. Corbin

Department of Crop Science
North Carolina State University
Raleigh, North Carolina 27695-7627

Thomas J. Monaco and Leslie A. Bjelk

Department of Horticultural Science
North Carolina State University
Raleigh, North Carolina 27695-7627

I. INTRODUCTION

The use of ^{14}C and other radioactive isotopes incorporated into organic molecules, where the positions of the labeled atoms are known with precision from the synthesis techniques used in preparing the compounds, has been an invaluable tool for research involved with determining the environmental fate of herbicides. Isotopes of the same element react similarly in chemical and physical processes, and very minute quantities of a radioisotope can be detected by observation of emitted radiation. Small amounts of radiolabeled herbicides added to nonradioactive materials can be used as tracers. ^{14}C-labeled herbicides move with the unlabeled material, react chemically the same, remain in the same proportion in any divisions of the compound, and their presence and amount can be detected with modern microprocessor-controlled nuclear counters. When herbicides are labeled with high specific activities, only low concentrations of the radioisotopes are needed for accurate measurements, concentrations at the ng level can be determined with accuracy and precision, and the radiotracer is not phytotoxic. Beta particles are the most widely used tracers in the biological sciences since many of the biologically important elements are β-emitters. Examples include ^{14}C, ^{3}H, ^{32}P, ^{35}S, ^{38}Cl, ^{40}K and ^{131}I. The half-life of the tracer should be long enough so that only a small part of the tracer disappears during the experiment. However, in some instances it is useful to select isotopes with relatively short half-lives to assure that they become harmless quickly. For example, ^{38}Cl with a half-life of 37 minutes will be useful for several hours. After a time interval of five half-lives, only 3% of the original label will remain.

Radioassay techniques in herbicide research have been directed to many topics such as mechanisms of action (Secor and Cseke, 1988), soil adsorption, uptake by plant roots, translocation from sites of application to "sink" areas (Eastin, 1986), and metabolism of compounds to transformation products (Frear and Swanson, 1975). The radiolabeling of a wide range of functional groups for many classes of herbicides has provided an excellent resource for the identification of metabolic intermediate compounds (Frear *et al.*, 1983).

Since their discovery, the primary means of detecting radioactive substances included the electroscope, the cloud chamber, the Geiger–Müller tube, fluorescent screens, scintillation phosphors, and photographic film. In recent years, modern instruments such as imaging proportional counters (IPC) with computerized controls and data processors have provided researchers with rapid and efficient analyses. The procedures described in this chapter have been used by our graduate students with a general knowledge of chemistry and instrumental analysis. A sufficient introduction to recent developments is given to challenge the student to learn the more sophisticated techniques of radiotracer methodology described in other chapters of this book.

II. LIQUID SCINTILLATION COUNTERS

At present, liquid scintillation spectrometers are considered to be among the most sensitive instruments in use for accurate quantitative analyses of small amounts of radioactive herbicides (Neame and Homewood, 1974). Although many instruments can be programmed to give automatic printouts of disintegrations per minute (DPM), many biological samples of herbicides contain chemicals and pigments that interfere with transmission of light to photomultiplier tubes. Counts of experimental samples should be compared with known standards and with regression equations from quench correction plots to assure the accuracy of measurement.

III. IMAGING PROPORTIONAL COUNTERS

A. Introduction

We have used thin-layer chromatography (TLC) for many years for the separation and identification of herbicide metabolites. Much of the early work on herbicide metabolism involved the time-consuming location of radioactive spots with autoradiography and TLC scanners. After locating the radioactive areas and verifying the cochromatography of ^{14}C samples with known standards, the ^{14}C zones were scraped into separate vials and quantified by liquid scintillation spectrometry. In recent years, substantial improvements have been made for the quantitative analysis of radiotracers in TLC through the development of computer controlled imaging proportional counters. IPC has been developed to give high sensitivity digital data from an entire TLC separation in one measurement. Within one minute both qualitative and quantitative evaluations can be achieved. The imaging capability provides a 100% improvement in sensitivity over mechanical scanners. Even with low tritium efficiencies of 1–2%, a sample can be quantified in ten minutes with 500 DPM of label. Sensitivity is 100 DPM or less with ^{14}C and higher energy isotopes (Shulman, 1983; Rock *et al.*, 1988).

B. Instrument Calibration

An instruction manual is provided by the manufacturer for the initial instrument settings. Also, a modify command is included with the computer control to assure alignment of the TLC plate origin, the 0-cm mark, and the 20-cm position on the plate. A standardized reference plate should be used frequently to assure correct listings of Rf values and to calibrate the actual position of radioactive spots on the plate with the plotted position of the printed scan.

1. Equipment
- Imaging proportional counter
- Computer and printer
- Standardized reference plate
- TLC plate with spots of known ^{14}C
- Liquid scintillation spectrometer

2. Procedure

a. Position the standardized reference plate on the scale calibration lane.
b. Collect counts for 1 minute.
c. Compare the actual position of the spots on the reference plate with the plotted position of the peaks on the printed scan.
d. Adjust the instrument in small increments (0.01 to 0.02 per increment).
e. Adjust the position to the left or to the right to assure that the 0-cm mark is aligned precisely with the edge of the plate (Fig. 1).
f. Count known amounts of ^{14}C on an experimental TLC plate to determine counting efficiency. Compare the counts from IPC with equivalent counts from a liquid scintillation counter (LSC) to obtain the specific activity (Fig. 2).

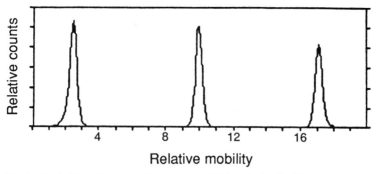

Fig. 1 Typical imaging proportional counter (IPC) analysis of calibration standard.

Absolute radioactivities can be obtained by multiplying the percent of total radioactivity from IPC by the absolute total radioactivity obtained by LSC analysis of a liquid sample.

C. Metabolism of Metribuzin in Cell Suspension Cultures

In recent years we have used cell suspension cultures as an adjunct to whole plant studies on the metabolism of herbicides. This is a rapid technique for metabolism studies and the first metabolite in a degradation series can often be extracted and identified without interference from other chemicals.

1. Chemicals and Equipment

NaOCl	Gamborg B5 media
Agar	^{14}C-metribuzin
Petri dishes	Rotary shaker
Tissue homogenizer	TLC plates (Silica gel HF)
Liquid scintillation spectrometer	Imaging proportional counter

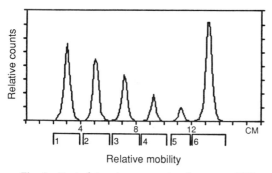

Fig. 2 Typical imaging proportional counter (IPC) analysis of analytical standards of a herbicide.

2. Procedure

a. Sterilize seeds in 20% commercial chlorine bleach containing 5% sodium hypochlorite for 10 minutes.

b. Wash in sterile distilled water and place sterile seeds on callus media solidified with 0.9% (w/v) agar.

c. After germination of seeds, transfer sections of radicles or hypocotyls to petri plates of Gamborg B5 media and make bimonthly transfers of calli to fresh media.

d. Place calli in Gamborg B5 liquid media after four bimonthly transfers on solid media. Grow cell suspensions on a rotary shaker (110 rpm) at 26°C.

e. Treat cell suspensions with ^{14}C-metribuzin for 1, 2, and 4 days.

f. Harvest the cells by filtering and washing with distilled water.

g. Weigh the washed cells, homogenize in a tissue grinder, and quantify ^{14}C in aliquots of the filtrate by liquid scintillation spectrometry.

h. Evaporate extracts of the cells and culture media under vacuum and spot 200 μl of the evaporated extracts on TLC plates. Develop to a 15-cm solvent front in benzene, acetone (2:1, v/v).

i. Identify positions of radioactive metabolites with an imaging proportional counter and quantify the amount of ^{14}C in each peak.

This experiment was conducted to demonstrate the tolerance of a tetraploid soybean line and the sensitivity of a diploid line. The tetraploid cells metabolized the ^{14}C-metrabuzin to the nontoxic deaminated metabolite. By contrast, the lack of metabolism of the herbicide resulted in death of diploid cells.

D. Herbicidal Inhibition of Acetyl-CoA Carboxylase

Two novel areas of herbicide chemistry that selectively provide postemergence control of monocot weeds in broadleaf crops are the aryloxyphenoxypropionic acid and the cyclohexanedione herbicides. The site of action of these two classes of herbicides has been shown to be inhibition of acetyl coenzyme A carboxylase, the first committed step in fatty acid biosynthesis. The purpose of this study has been to isolate partially purified acetyl coenzyme A carboxylase (ACCase) and to measure herbicide selectivity as a function of differential sensitivity at the site of action in rice and monocot weeds. The partially purified protein (containing ACCase) is assayed with ATP, acetyl CoA, NaH^{14}CO$_3$ and various dosages of the herbicidal inhibitors to detect a decreased formation of ^{14}C-malonyl CoA from the substrates.

1. Chemicals and Equipment

NaH¹⁴CO₃	Formic acid
Acetyl CoA carboxylase	Ether
KOH	HCl
Malonyl CoA	ITLC-SG plates
Malonic acid	

2. Procedures

a. Isolate and partially purify acetyl coenzyme A carboxylase according to the techniques of Secor and Cseke (1988) and Thomson and Zalik (1981).

b. Precipitate the proteins with $(NH_4)_2SO_4$, which is a modification of the above techniques.

c. Dilute the isolated protein to approximately 0.1 mg/ml to use with the substrates to measure acetyl CoA-dependent incorporation of NaH¹⁴CO₃.

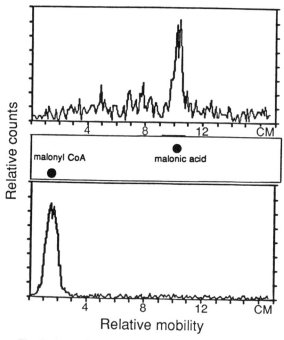

Fig. 3 Typical imaging proportional counter (IPC) analysis of hydrolyzed (Rf 0.92) and unhydrolyzed (Rf 0.01) end-product of NaH¹⁴CO₃ incorporation in the presence of acetyl-CoA, partially purified ACCase, and ATP. The solvent system is water-saturated ether, formic acid (7:1, v/v).

d. Terminate the reaction with 1N HCl and hydrolyze the thio-ester of malonyl CoA in 3N KOH for 10 min at 35°C.

e. Neutralize the solution to pH 7.0.

f. Spot aliquots of labeled hydrolyzed and unhydrolyzed end-products and unlabeled standards of malonyl CoA and malonic acid on Gelman ITLC-SG plates.

g. Develop the plates in a solvent solution of water-saturated ether, formic acid (7 : 1, v/v).

h. Locate the unlabeled malonyl CoA and malonic acid by spraying the plates with a slightly alkaline solution of 0.1% bromcresol green.

i. Locate and quantify the radioactive end-products with IPC.

j. Compare the location of each radioactive peak with the location of the unlabeled standards.

This experiment demonstrates how the end product of the reaction was chromatographed and measured with IPC to detect whether the end product of the reaction is malonyl CoA, the first product in fatty acid biosynthesis. Figure 3 illustrates the results with IPC of the labeled end-product versus the unlabeled standards. The hydrolyzed end-product cochromatographed with the unlabeled malonic acid near the solvent front. The unhydrolyzed radioactive end-product remained at the origin with the unlabeled malonyl CoA standard. This study demonstrates that the end product of the reaction of partially purified acetyl coenzyme A carboxylase, ATP, CO_2, and acetyl CoA is malonyl CoA and provides evidence for location of the site of action of the herbicides (Burton *et al.*, 1987, and Secor and Cseke, 1988).

REFERENCES

Burton, J. D., Gronwald, J. W., Somers, D. A., Connelly, J. A., Gengenbach, B. G., and Wyse, D. L. (1987). Inhibition of plant acetyl coenzyme A carboxylase by the herbicides sethoxydim and haloxyfop. *Biochem. and Biophys. Res. Comm.* **148**, 1039–1044.

Eastin, E. F. (1986). Absorption, translocation, and degradation of herbicides by plants. In "Research Methods in Weed Science" (N. D. Camper, ed.), pp. 277–289. Southern Weed Science Society, Champaign, IL.

Frear, D. S., and Swanson, H. R. (1975). Metabolism of cisanilide by excised leaves and cell suspension cultures of carrot and cotton. *Pestic. Biochem. Physiol.* **5**, 73–80.

Frear, D. S., Swanson, H. R., and Mansager, E. R. (1983). Acifluorfen metabolism in soybean:diphenylether bond cleavage and the formation of homoglutathione, cysteine, and glucose conjugates. *Pestic. Biochem. Physiol.* **20**, 299–310.

Neame, K. D., and Homewood, C. A. (1974). "Liquid Scintillation Counting." John Wiley and Sons, New York.

Rock, C. O., Jackowski, S., and Shulman, S. D. (1988). Imaging scanners for radiolabeled thin-layer chromatography. *Biochromatography* **3**, 127–130.

Secor, J., and Cseke, C. (1988). Inhibition of acetyl CoA carboxylase activity by haloxyfop and tralkoxydim. *Plant Physiol.* **86,** 10–12.

Shulman, S. D. (1983). A review of radiochromatogram analysis instrumentation. *J. Liquid Chrom.* **6,** 35–53.

Thomson, L. W., and Zalik, S. W. (1981). Acetyl coenzyme A carboxylase activity in developing seedlings and chloroplasts of barley and its virescens mutant. *Plant Physiol.* **67,** 661–665.

7

Aquatic Toxicology: Degradation of Organic Xenobiotics

A. E. McElroy

Environmental Sciences Program
University of Massachusetts-Boston
Boston, Massachusetts 02125

I. INTRODUCTION

Pollution of the environment with xenobiotic organic compounds has been a topic of intense scientific investigation for almost 20 years. With the realization that a seemingly innocuous compound such as the pesticide DDT had rapidly become distributed worldwide and was causing reproductive failure in nontarget species (as publicized by Rachel Carson, 1962), the need to understand the environmental fate and effects of xenobiotic contaminants became abundantly clear. Although persistent, many organic contaminants are subject to biotic and abiotic transformations that yield products with altered physical/chemical and biological properties. Therefore, any thorough investigation of environmental fate and effects must include analysis of both the parent compound and any transformation products that may be produced.

Traditional analytical chemical methodology employs chromatographic cleanup procedures designed to remove interfering chemicals with altered polarity or size, which effectively removes most transformation products. Furthermore, many of the metabolites of persistent xenobiotics are relatively nonvolatile and of high enough molecular weights to disallow analysis by standard gas chromatography and mass spectrometry. New techniques are currently being developed to alleviate these problems, but they are still highly experimental. Most require sophisticated equipment to couple liquid chromatographs to mass spectrometers and/or fast-atom bombardment ionization chambers, or are indirect measures of biochemical modification,

CARBON ISOTOPE TECHNIQUES

109

such as using ^{32}P to postlabel metabolites adducted to DNA, or using antibodies to detect the induction oɪ specific isozymes responsible for metabolism. At the present time, use of radioactive carbon isotopes to trace the environmental fate of xenobiotics is still a powerful tool because: (1) ^{14}C provides the increased sensitivity needed to trace organic contaminants in relatively small samples of many different types; and (2) ^{14}C allows unambiguous quantification of unknown transformation products. Because only the radioactive signal is followed, extensive sample cleanup is not necessary; therefore relatively large numbers of samples can be processed rapidly at reasonable cost with equipment generally available in most laboratories.

Although radioactive tracers are frequently used as research tools in aquatic toxicology, there are no generally accepted standard methods or protocols. A general overview of some of the basic concepts in experimental design can be found in Anderson and Conning (1988). Investigators develop analytical methodologies appropriate for their particular system. In part this is predicated by the differing physical/chemical properties of the compounds under investigation. As long as specific methods are thoroughly described and verified, any number of approaches can be used. Unfortunately, the multiplicity of methods commonly used makes comparisons of results difficult in some cases.

The primary experience of the author has been with the environmental fate of persistent organics such as polycyclic aromatic hydrocarbons (PAH) and polychlorinated biphenyls (PCBs) in aquatic ecosystems. Methodology developed to study these systems will be used to illustrate approaches that can be used in aquatic toxicology of hydrophobic organic contaminants. General procedures will be discussed for determination of total radioactivity and metabolic transformation products in environmental samples. Specific methods used for the analysis of metabolites of PAH in animal tissues and sediment samples will then be discussed. Finally, considerations important to the use of these methods to answer questions about environmental fate and metabolism in microcosm experiments and within individual aquatic organisms will also be discussed.

II. MATERIALS REQUIRED

Materials needed for using ^{14}C in environmental toxicology are generally the same as those needed in any ^{14}C study as outlined in Chapter 1, this volume. In addition, any work with live aquatic organisms will require an exposure system minimally consisting of an uncontaminated source of either fresh or salt water, and a charcoal filtration system to trap either contaminants in effluent from flow-through systems or material excreted by organisms in recirculating systems. A secondary containment system large enough to keep any spills due to tank overflow or failure from contaminating the rest of the work area is strongly advised. Whenever possible, all aquaria and labware should be made of glass, stainless steel, or solvent-inert plastics such as Teflon to minimize sorption of organic contaminants and allow strong solvents such as hexane, acetone, or chloroform to be used for decontamination or extraction.

Extraction and separation of environmental contaminants requires glassware and equipment generally found in any organic or biochemistry laboratory such as a variety of organic solvents, homogenization equipment, some means of reducing the volume of extracts (such as a rotary evaporator, speed-vac, or distillation setup), a table-top centrifuge, and a small- to medium-scale filtration system. Once free from the sample matrix, biotransformation products will need to be separated using some form of chromatography and the products quantified and identified. As discussed below, the equipment used in this stage of the process can range from relatively inexpensive disposable cartridges coupled with standard liquid scintillation counting (LSC) to relatively expensive high pressure liquid chromatographs with on-line radioactivity detectors, depending on what resolution is required and budgetary constraints.

III. DESCRIPTION OF PROCEDURES

A. General Methods

1. Total Radioactivity

Total radioactivity in up to several ml of an aqueous sample can be determined directly using standard LSC techniques. Radioactivity incorporated into sample matrices such as biological tissue or sediment needs to be released prior to LSC. Radioactivity in tissues can be liberated using either a commercially available tissue solubilizer or a strong $(1-2N)$ base solution. Most commercial tissue solubilizers are quaternary amine derivatives. Amounts needed vary with application but generally 0.4 g of wet tissue can be dissolved in $1-2$ ml of solubilizer. Mild heating at $50°C$ and gentle agitation will speed digestion, which should be done in vials with Teflon or polyethylene cap liners, as the solubilizers will dissolve foil cap liners. Manufacturers of tissue solubilizers provide specific instructions on the use of their products. Due to the high cost of commercial solubilizers, when sufficient, use of strong base is recommended. Regardless of alkali digestion method used, care should be taken to decolorize (using hydrogen peroxide or sodium hypochlorite) and to neutralize (using either concentrated acid and/or a strong buffer such as Tris) prior to LSC. Tissues that are resistant to digestion such as caraspace, bone, or cuticle can sometimes be dissolved using nitric or 70% perchloric acid at 70°C. Specific descriptions of these methods can be found in Peng (1981).

For determination of total radioactivity in sediment samples or tissues resistant to solubilization or the oxidation technique described above, wet oxidation or dry combustion of samples to $^{14}CO_2$ and subsequent collection of CO_2 in an appropriate base is recommended. Coughtrey *et al.* (1986) describe a wet oxidation technique using potassium dichromate and concentrated sulphuric and phosphoric acid, which can be done in a modified filter flask. This technique can accommodate up to 0.3 g of dry biological material or up to 5 g of dry soil. Cheng and Farrow (1976) describe a dry combustion technique using a modified LECO high-temperature induction furnace. Considering the explosive nature of the reactants (the samples are combusted in an oxygen atmosphere), great care should be exercised when designing such a system. Commercial manual or automatic sample oxidizers can be obtained from Packard (Downers Grove, Illinois) or Harvey Instruments (Hillsdale, New Jersey), which can accommodate up to 0.3 g of biological material. However the cost of commercial oxidizers is considerable ($12,000 to $26,000, depending on model and options). Regardless of the method utilized to determine total radioactivity, nonradioactive samples of similar consistency should be spiked with known amounts of isotope to determine recovery efficiencies and correct for quench and chemiluminescence.

2. Metabolite Analysis

a. Extraction In order to separate, quantify and ultimately identify metabolic products, it is first usually necessary to extract the parent compound and metabolic products from tissue, sediment, or water samples. The method of extraction chosen will entirely determine the range of compounds that can be examined. Unfortunately there are no "standard" techniques in this field. There are four major classes of compounds that need to be considered: (1) the unchanged parent compound; (2) primary metabolites (those that have been made more soluble by the addition of a functional group, but that can still be extracted into organic solvents); (3) secondary metabolites or conjugates (metabolites that have been further modified to form water-soluble compounds); and (4) bound residues (covalently and noncovalently bound metabolites that are recalcitrant to extraction in either organic or aqueous media). Figure 1 shows some of the metabolites that can be formed from the PAH benz(a)anthracene. The complexity of the extraction protocol will depend on how many of these different classes the investigator wishes to examine.

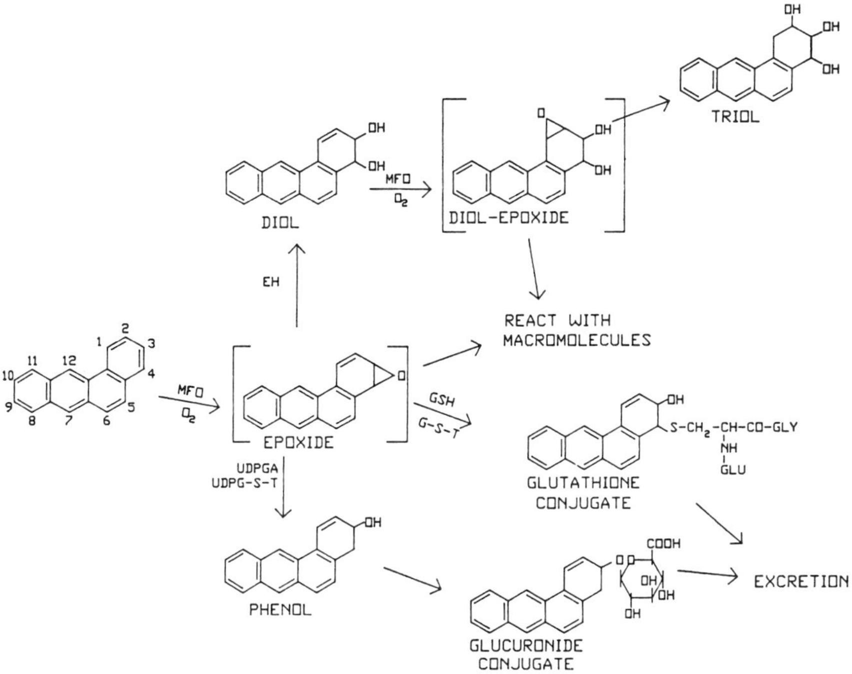

Fig. 1 Metabolites of benz(a)anthracene: MFO, Mixed function oxygenase; G-S-T, gluta-thione S-transferase; GSH, reduced glutathione; UDPGA, uridine diphosphate glucuronic acid; UDPG-S-T, uridine diphosphate glucuronosyl S-transferase; Gly, glutamate; Gly, glycine.

Since most organic contaminants are lipid soluble, methods have been adapted from those used for the analysis of pesticides. Protocols employing acetone, methanol, ethyl acetate, acetonitrile, methylene chloride, chloro-form, and hexane alone and in combination for extraction are frequently encountered. When deciding on an extraction protocol one must make sure that there are no physical or chemical barriers to keep the extraction solvent from freely associating with the analytes of interest. This usually involves homogenization of the sample either before or during extraction, and ensur-ing that the sample matrix is miscible in the extraction solvent. Grinding the sample in an anhydrous coarse salt such as sodium sulfate prior to extraction helps to macerate the sample, solves potential problems of solvent misci-bility (because the salt absorbs all available water), and allows quantitative subsampling of homogenates for either replicate or alternative extraction protocols. If only organic soluble metabolites are of interest, relatively simple extraction schemes using multiple extractions with acetone, acetoni-

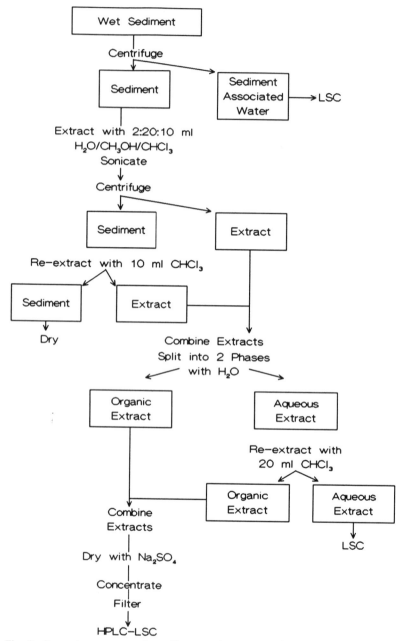

Fig. 2 Extraction sequence for sediment and tissue samples: LSC, liquid scintillation counting; H₂O, water; CH₃OH, methanol; CHCl₃, chloroform; Na₂SO₄, sodium sulfate; HPLC, high-performance liquid chromatography.   (*Continues*)

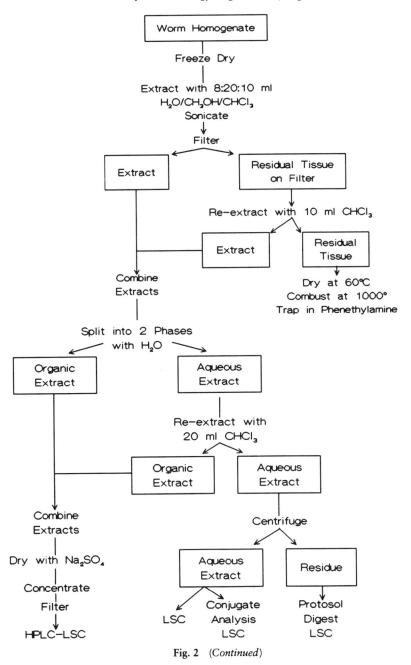

Fig. 2   (*Continued*)

trile, ethyl acetate, or methanol may be adequate. However, if water-soluble or conjugated metabolites are of interest, a more detailed procedure must be used.

A broad spectrum general lipid extraction scheme modified after one first described by Bligh and Dyer (1959) is shown in Fig. 2. This scheme, developed for the PAH benz(a)anthracene, when carried to completion will quantify total radioactivity as well as that radioactivity remaining as parent compound, and that converted into individual polar metabolites, conjugate classes, and that which is unextractable or bound. The initial ratio of water, methanol, and chloroform is chosen to maintain a single phase mixture that can extract analytes with a wide range of solubility. Additional water and chloroform is added to re-extract the sample and split the extract into two phases, so that organic soluble and aqueous metabolites can be analyzed separately either as total radioactivity or after further separation as described below. In our laboratory, this extraction procedure has been found to be greater than 90% efficient at recovering unmetabolized parent compound.

It is important to determine any residual activity remaining after extraction. Such radioactivity can result from insufficient extraction methodologies or truly unextractable material such as that covalently bound to elements of the sample matrix. In both sediment and tissue samples this fraction can be considerable and is of toxicological importance. After exhaustive extraction, total residual radioactivity can be determined by either digesting, oxidizing, or combusting the residue. If significant counts are found, different extraction procedures can be used on fresh tissue to determine how much of the unextractable material is covalently bound to protein, DNA, or RNA. Von Hofe and Puffer (1986) describe a reasonably straightforward method using phenol extraction to remove protein-bound material followed by extraction of nucleosides in a urea buffer from which DNA and RNA can be recovered by chromatography with hydroxylapatite (Adriaenssens et al., 1982).

b. Separation Separation of the parent compound from polar but organic soluble metabolites is usually achieved using reverse phase high-performance liquid chromatography (HPLC), where more polar metabolites elute prior to the parent compound. Metabolite production is quantified using standard LSC on fractions collected throughout the run, with tentative identification of metabolites produced based on coelution with known standards detected by ultraviolet absorption or fluorescence. Standards for several PAH metabolites can be obtained with permission from the National Cancer Institute (Bethesda, Maryland). An example of this approach is shown in Fig. 3 taken from Stein et al. (1984) where bile from fish exposed to sediments containing PCBs and the PAH benzo(a)pyrene was chromatographed before and after

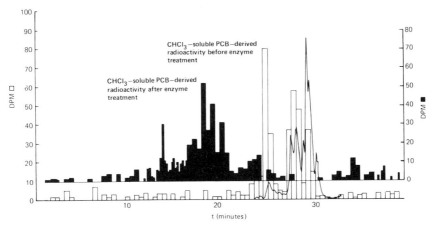

Fig. 3 HPLC radiochromatogram of PCB metabolites in fish bile before and after (shaded) treatment with aryl sulfatase and β-glucuronidase. Separation was achieved using a Perkin-Elmer HC ODS/SIL-X 0.26 × 25-cm column and a nonlinear gradient between (A) 99.5/0.5 (v/v) water/acetic acid and (B) methanol, starting at 80% A for 0.5 min, dropping to 50% A over 9 min, then to 30% A in 12 min, to 0% A in 10 min, and finally holding at 0% A for another 8 min at a flow rate of 1.0 ml/min and a column temperature of 35°C. Adapted from Stein *et al.* (1984).

enzymatic hydrolysis to cleave conjugates (see discussion below) using a nonlinear gradient of methanol and water.

Several on-line radioactivity monitors are commercially available, which either use a solid scintillant column or mix cocktail in with the elutrate from the HPLC prior to introduction to a flow-through photodetector to provide real-time radioactivity detection. Although very expensive ($10,000 to $22,000, depending on options and data-handling capabilities, using current prices from the Radiomatics division of Packard Inst., Tampa, Florida, for example), on-line detectors hold great promise to drastically cut down the time and expense and liquid waste disposal problems generated by collecting and counting hundreds to thousands of fractions per day using standard LSC techniques. However, because a relatively small amount of material is being counted for a relatively short period of time, as compared with standard LSC techniques, detection limits are usually higher. In addition, depending on the mobile phase and the compounds being separated, irreversible coating of the solid scintillant packing can rapidly increase background counts, thereby reducing sensitivity. A comparison of traditional fraction collection and LSC versus on-line radioactivity detection for quantification of compounds separated by reversed phase HPLC is presented by Kessler (1982).

Tissues with extremely high lipid contents such as crustacean hepatopancreas or fish liver can make clean separations of HPLC difficult due to the large amount of lipid extracted with the contaminant. If this is a problem, exhaustive extraction with a nonpolar solvent such as hexane to remove the parent compound only, followed by extraction in more polar solvents such as ethyl acetate or acetonitrile to release polar metabolites for chromatography may be employed.

If an HPLC is not available there are less expensive ways to separate the parent compound from polar metabolites. Either normal phase (silica) or reverse phase (C_{18}) thin layer chromatography (TLC) can be used to quickly determine whether or not metabolites are being produced. Separations on TLC plates can be quantified using autoradiography (incubating very sensitive photographic film over the plate in the dark for periods of days to weeks) or by transferring segments of the chromatogram into scintillation vials for LSC.

Simple separations can also be affected using mini-columns. A number of chromatographic packing materials including silica, florosil, and various reverse phase supports are available in disposable cartridges or columns. These can be used to concentrate analytes from large volumes of either aqueous solutions (reverse phase) or organic extracts (normal phase), to effect a partial cleanup of sample prior to more rigorous separations, or to separate parent compound from polar metabolites. With either TLC or disposable cartridges, relatively small number of fractions are generated, which can be counted using standard LSC methods.

If it is necessary only to separate the parent compound from all polar metabolites, liquid/liquid extraction methods can be used. In this case the organic extract is taken to dryness in a test tube. The sample is then equilibrated between two immiscible solvents, one in which only the parent compound is soluble, and one which should extract all metabolites. For the PAH benzo(a)pyrene, multiple extraction with 2.0 ml *n*-hexane against 1.0 ml 0.15 *M* KOH in 85% dimethylsulfoxide (DMSO) works well. This technique was adapted from the benzo(a)pyrene hydroxylase assay developed by De Pierre *et al.* (1975) and modified by Van Cantfort *et al.* (1975). After vortexing the sample in the solvents, they can be cleanly separated by a several-minute spin in a desk-top centrifuge. The hexane fraction is then pipetted off the aqueous layer, the extraction repeated, and the hexane fractions combined. After solvent reduction, two fractions remain, one containing the unmetabolized benzo(a)pyrene and the other containing any metabolites which are then counted with LSC. We have found that 5 extractions are sufficient to remove >98% of unmetabolized benzo(a)pyrene.

Conjugated (water-soluble) metabolites can be further identified in a number of ways. Water-soluble extracts can be treated with aryl sulfatase and or β-glucuronidase to cleave either sulfate and/or glucuronide conjugates (see Stein *et al.*, 1984, for one approach). If determining the type of conjugate is not important, incubation in 2N acid will also cleave sulfate and glucuronide conjugate bonds. Once the conjugates have been liberated, the mixture is then re-extracted with organic solvents. Any conjugate liberated by enzymatic treatment would then be converted back into the polar metabolite from which it was formed, and could then be identified by using chromatography (see Fig. 3). In most cases the presence of glutathione conjugates is inferred by difference (after accounting for sulfate and glucuronide conjugates). Recently, HPLC methods have been developed to separate sulfate, glucuronide, and glutathione conjugates without prior enzymatic cleavage using a C_8 column and gradient elution with increasing strength of methanol versus a pH 7.4 0.04 M tetrabutylammonium bromide buffer (Smolarek *et al.*, 1987).

Knowing the physical chemical properties of the parent compound and potential metabolites, it should be possible to design an extraction and chromatographic protocol to isolate and separate the compounds of interest. Care must also be taken to avoid unintentional alteration during experimentation or analysis. Some PAH, and particularly their polar metabolites, are extremely sensitive to photooxidation. If photolysis is not meant to be part of the experiment, exposure to ultraviolet light should be avoided. Particularly when the compounds of interest are dissolved in solution, they should be protected from direct light. Some investigators conduct metabolism work with PAH entirely under red or gold illumination to minimize photoreactions. It is always important to verify the extraction efficiency and separations attempted using known standards added to unlabeled samples. Unfortunately, reference standards incorporated into environmental matrices such as animals' tissue or sediment are not available for metabolic products. Tentative identities of metabolites detected using ^{14}C can be made based on optical spectra or coelution with known pure standards; however, confirmation should be done with mass spectrometry and/or nuclear magnetic resonance spectroscopy.

Regardless of which of the approaches described above are employed, before attempting to assess metabolism it is necessary to ensure that the radiolabeled parent compound is pure. Many commercially available isotopes can be obtained in a relatively pure state (>98%). However, purity should always be checked to ensure that subsequent physical or chemical alteration has not occurred. It is also important to run poisoned controls, or at least run blanks spiked with the starting material throughout the entire analytical scheme to assess potential procedural (i.e., nonmetabolic) alteration of the compound. Compound purity is most frequently checked using

TLC or HPLC chromatography. If degradation is observed, the compound can be repurified using chromatography or any of the extraction procedures described above.

B. Microcosm Work

Radiolabeled contaminants have been successfully used in microcosms to determine the environmental fate and effects of contaminants and their metabolites. When attempting work of this sort, the design of the exposure system and method of administering the contaminant need to be as environmentally realistic as possible. Because of the hydrophobic nature of many organic contaminants, interactions with the benthos are important in experimental design, and solubility constraints must be considered in determining the method of dosing. In complex systems, construction of a mass balance of the fate of all added radiolabel can be integral to the successful characterization of key processes and reservoirs.

Figure 4 shows small (2 l) and large (228 l) microcosms that have been used by the author to determine the fate of radiolabeled PAH benz(a)anthracene and its metabolites in benthic systems. The small flow-through microcosms were designed to allow repeated sampling of sediment and water column with time. Radioactivity in the water column was collected for determination of total extractable radioactivity (after concentration onto disposable C_{18} cartridges), or could be rerouted into a second chamber for determination of $^{14}CO_2$ produced (McElroy, 1985). Similar measurements were made using the 228-l system (Fig. 4), which recirculates seawater over a 0.37-m^2 sediment box (McElroy et al., 1987). Experiments done in these microcosms and the even larger mesocosms at the Marine Ecosystem Research Laboratory (MERL) at the University of Rhode Island (Lee et al., 1982) have demonstrated that PAH such as benz(a)anthracene can be rapidly metabolized in benthic ecosystems, and that under some circumstances metabolic products persist for periods of weeks to months.

Because of their negligible water solubility (≤ 10 ppb), when working with environmental contaminants such as the more highly chlorinated PCBs or some of the larger (> 3 ring) PAH, sediment, particulate, or generator column dosing systems provide the best method of introduction. Sediment can be labeled by adding the isotope to a sediment slurry in a small volume of solvent and then equilibrating the slurry with constant mixing until the sediment is uniformly labeled. Another approach is to coat the sides of the labeling vessel with the isotope in an appropriate solvent, allow the solvent to evaporate, and then equilibrate the sediment slurry with the coated container wall. This approach avoids potential artifacts due to influence of the carrier solvent on the behavior or the effect of the contaminant. Experiments labeling a high and low carbon (4 vs. 2% organic carbon) fine sedi-

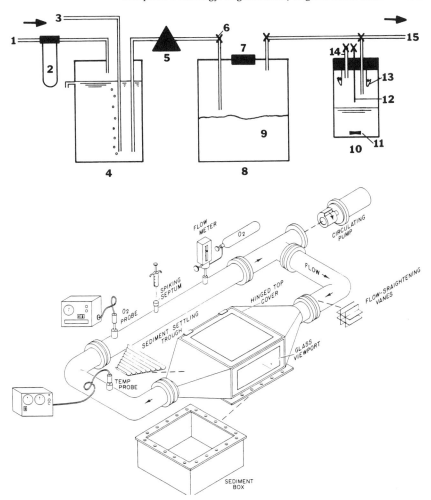

Fig. 4 Small and large microcosms for the study of PAH metabolism in benthic systems. Small benthic chamber: 2-l total chamber volume, flow-through, all glass construction; 1, seawater inlet; 2, 1-μm honeycomb filter; 3, air line; 4, seawater reservoir; 5, peristaltic pump; 6, 3-way stopcock; 7, silicon stopper; 8, exposure chamber; 9, sediment; 10, CO_2-trapping chamber; 11, magnetic stirring bar; 12, acid injection port; 13, base-trapping wicks; 14, vent; 15, waste. Large benthic chamber: 228 l recirculating fluid volume. Adapted from McElroy *et al.* (1987).

ment with ^{14}C-hexachlorobiphenyl demonstrated that bioavailability of the labeled contaminant was not altered when the equilibration period was extended from one to four weeks (McElroy and Means, 1988). Ray and Giam (1987) used a similar technique to coat hydrophobic organics onto sand grains. After evaporation of solvent, they could produce consistent low levels of dissolved hydrophobic organics by maintaining a constant flow of water over a column of labeled sand that was suitable for bioaccumulation studies in fish. Pruell et al. (1986) have developed a system that maintains a constant load of suspended particulates, which has been successfully used to expose filter feeding bivalves to organic contaminants.

C. *In Vivo* Metabolism

From an environmental perspective *in vivo* metabolism studies represent a subset of the microcosm experiments where the organism becomes the system. As above, designing experiments so that a mass balance of the fate of the radiolabel is obtained can yield valuable information that might be missed if only selected compartments are analyzed. Efforts should be made to accurately quantify the exact amounts and form of the compound reaching the organism, what are the sites of metabolism and deposition within the body, and what are the routes and forms of excreted material.

As in microcosms, route of exposure is important. Much toxicokinetic work has been done using either intravenous (IV) or intraperitoneal (IP) injections of toxicant. Although these routes are relevant in the study of drugs, in the environment organisms rarely encounter toxicants in this form. Intravenous dosing will give an estimate of the fate of a completely absorbed substance once it reaches the blood stream. However, for compounds that may undergo biotransformation during absorption through the gastrointestinal tract or the gills, disposition of an intravenous dose may not generate environmentally relevant information. For pelagic organisms or filter feeders, presentation of the toxicant in the dissolved or particulate state or through the diet would be most environmentally relevant. For benthic organisms, exposure to contaminated sediments or contaminated diets would be most appropriate. Acute or single doses of contaminants are most often investigated, but a more realistic environmental exposure scenario would be either repeated or continuous dosing. Particularly where metabolism is concerned, prior or continuous exposure to some toxicants can alter the metabolic capacity of the organism (see reviews in Varanasi, 1989). In the case of continuous exposure, organisms should be followed until they reach an apparent steady state of accumulation and then placed in an uncontaminated system to assess depuration.

Sampling should be timed to determine initial uptake into the blood or hemolymph, distribution to key organs, and in the case of single or acute dosing, the short- and long-term disposition of radiolabel remaining in the body. Such an approach usually requires repeated sampling of the distribution fluids during the first hours after exposure, and determination of the tissue distribution of parent compound and metabolites in individuals sacrificed at periods of days to weeks. When possible, indwelling catheters should be used to collect uncontaminated and repeated samples of blood, urine, and fecal material. When cannulation is not possible, flow-through exposure systems should be used and the tank effluent monitored for total radioactivity and the release of metabolic products. Methods developed for the study of drug deposition and metabolism in mammalian species can be adapted with care for use in aquatic species (see specific chapters in Illing, 1989).

IV. COMMENTS

This chapter was intended to present the reader with a few specific methods and some general guidelines that should be considered when using carbon isotopes to study aquatic toxicology and the degradation of organic xenobiotics. Considering the breadth of the subject, this introduction is of necessity just an overview. Anyone seriously considering conducting research of this type should carefully evaluate the questions they wish to address, consult the recent literature on specific methods and systems others have used to explore similar compounds, and then using appropriate standards, validate the approach chosen in their own particular system and laboratory. As more is learned about the fate of xenobiotics and their breakdown products in the environment, standard methods will begin to emerge. Only then can truly comparable data on many different systems and in many different organisms be generated. Until significant advances are made in contaminant analysis, radioisotopic tracer methodology will continue to be the method of choice.

REFERENCES

Adriaenssens, P., Bixler, D., and Anderson, M. (1982). Isolation and quantitation of DNA-bound benzo[a]pyrene metabolites: comparison of hydroxylapatite and precipitation procedures. *Anal. Biochem.* **123**, 162–169.

Anderson, D., and Conning, D. M. (1988). "Experimental Toxicology, the Basic Principles." Royal Society of Chemistry, London.

Bligh, E. G., and Dyer, W. J. (1959). A rapid method of total lipid extraction and purification. *Can. J. Biochem. Physiol.* **37**, 911–917.

Carson, R. (1962). "Silent Spring." Houghton Mifflin, Boston.

Cheng, H. H., and Farrow, F. O. (1976). Determination of ^{14}C-labeled pesticides in soils by a dry combustion technique. *Soil Sci. Soc. Am. J.* **40**, 148–150.

Coughtrey, P. J., Nancarrow, D. J., and Jackson, D. (1986). Extraction of carbon-14 from biological samples by wet oxidation. *Commun. in Soil Sci. Plant Anal.* **17**, 393–399.

De Pierre, J. W., Moron, M. S., Johannesen, K. A. M., and Ernster, L. (1975). A reliable, sensitive, and convenient radioactive assay for benzpyrene monooxygenase. *Anal. Biochem.* **63**, 470–484.

Illing, H. P. A. (1989). "Xenobiotic Metabolism and Disposition: The Design of Studies on Novel Compounds." CRC Press, Boca Raton, FL.

Kessler, M. J. (1982). Quantification of radiolabeled compounds eluting form the HPLC system. *J. Chromatog. Sci.* **20**, 523–527.

Lee, R. F., Hinga, K., and Almquist, G. (1982). Fate of radiolabeled polycyclic aromatic hydrocarbons and pentachlorophenol in enclosed marine ecosystems. In "Marine Mesocosms, Biological and Chemical Research in Experimental Ecosystems." (G. D. Grice and M. R. Reeves, eds.), pp. 123–135. Springer-Verlag, New York.

McElroy, A. E. (1985). Physiological and biochemical effects of the polycyclic aromatic hydrocarbon benz(a)anthracene on the deposit feeding polychaete *Nereis virens*. In "Marine Biology of Polar Regions and Effects of Stress on Marine Organisms." (J. S. Gray and M. E. Christiansen, eds.), pp 527–543. John Wiley & Sons, Ltd. New York.

McElroy, A. E., and Means, J. C. (1988). Factors affecting the bioavailability of hexachlorobiphenyls to benthic organisms. In "Aquatic Toxicology and Hazard Assessment: 10th Volume, ASTM STP 971." (W. J. Adams, G. A. Chapman, and W. G. Landis, eds.), pp. 149–158. American Society for Testing Materials, Philadelphia, PA.

McElroy, A. E., Tripp, B. W., Farrington, J. W., and Teal, J. M. (1987). Biogeochemistry of benz(a)anthracene at the sediment–water interface. *Chemosphere* **16**, 2429–2440.

Peng, C. T. (1981). "Sample Preparation in Liquid Scintillation Counting." Amersham Corp., Arlington Heights, IL.

Pruell, R. J., Lake, J. L., Davis, W. R., and Quinn, J. G. (1986). Uptake and depuration of organic contaminants by blue mussels, *Mytilus edulis*, exposed to environmentally contaminated sediment. *Mar. Biol.* **91**, 497–507.

Ray, L. E., and Giam, C. S., (1987). Bioaccumulation and depuration of selected organic compounds in marine fish. In "Pollutant Studies in Marine Animals." (C. S. Giam and L. E. Ray, eds.), pp. 23–50. CRC Press, Boca Raton, FL.

Smolarek, T. A., Morgan, S. L., Moynihan, D. G., Lee, H., Harvey, R. G., and Baird, W. M. (1987). Metabolism and DNA adduct formation of benzo[a]pyrene and 7,12-dimethylbenz[a]anthracene in fish cell lines in culture. *Carcinogenesis* **8**, 1501–1509.

Stein, J. E., Hom, T., and Varanasi, U. (1984). Simultaneous exposure of English sole *(Parophrys vetulus)* to sediment-associated xenobiotics: Part I—uptake and disposition of ^{14}C-polychlorinated biphenyls and ^{3}H-benzo[a]pyrene. *Mar. Environ. Res.* **13**, 97–119.

Van Cantfort, J., De Graeve, J., and Gielen, J. E. (1975). Radioactive assay for aryl hydrocarbon hydroxylase improved method and biological importance. *Biochem. Biophys. Res. Comm.* **79**, 505–512.

Varanasi, U. (1989). "Metabolism of Polycyclic Aromatic Hydrocarbons in the Aquatic Environment." CRC Press, Boca Raton, FL.

Von Hofe, E., and Puffer, H. W. (1986). *In vitro* metabolism and *in vivo* binding of benzo(a)pyrene in the California killifish *(Fundulus parvipinnis)* and speckled sanddab *(Citharicthys stigmaeous)*. *Arch. Environ. Contam. Toxicol.* **15**, 251–256.

8

Carbon Dating

K. M. Goh

Department of Soil Science
Lincoln University
Canterbury, New Zealand

I. INTRODUCTION

The ubiquitous presence of ^{14}C in biological materials (e.g., wood, peat, soils, bones, shells) enables carbon to be dated and thereby provides an estimate of the ages of biological systems. The time frame of the ^{14}C dating technique from 200 to 40,000 years before the present (1950) spans a period of important global environmental changes, thus making ^{14}C dating one of the most useful techniques for studying Quaternary climatic fluctuations (Fig. 1). However, a number of assumptions are implicit in the ^{14}C dating method. It is vital that users of this technique understand their significance and adopt procedures to overcome some of the limitations to improve the reliability of ^{14}C dates.

II. REQUIREMENTS OF CARBON DATING

Due to the low ^{14}C activity (e.g., it is estimated that there is 1 atom of ^{14}C in about 10^{12} atoms of ^{12}C) commonly present in biological materials submitted for ^{14}C dating that have not been enriched with "modern" or "bomb" ^{14}C resulting from the detonation of thermonuclear devices in the 1950s and 1960s (Goh *et al.*, 1977b), extremely sensitive equipment is required to detect and count ^{14}C emissions. In practice, ^{14}C measurements are conducted in ^{14}C dating laboratories by chemists and physicists whose task is to

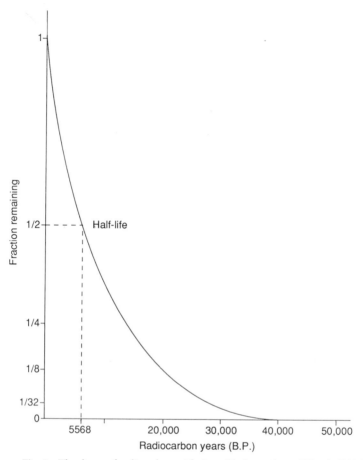

Fig. 1 The decay of radiocarbon with time. [Redrawn from Gillespie (1984). Reproduced from p. 1 by permission of Oxford University Committee for Archaeology.]

ensure as accurate a ^{14}C date as can be obtained from the sample received. The collection and often the preparation of samples are the responsibility of the field workers or biologists known as ^{14}C users.

A.   Requirements of Carbon Dating Laboratories

One of the main responsibilities of ^{14}C dating laboratories is to ensure that the precision of ^{14}C counting with the equipment available is maximized to improve the accuracy of dating. This is achieved by being familiar with the function of a particular counter, the sample size required, length of counting time needed, and knowledge of background and modern standards, instrument consistency, and corrections of isotopic fractionation.

Some laboratories conduct physical and chemical pretreatments of samples to remove contaminants before dating. Most laboratories require samples to be submitted in a sufficiently large amount to meet the counting instrument requirement and in a state that requires minimum laboratory pretreatments. These are best sorted out before commencing the collection of samples for dating.

B.   Requirements of ¹⁴C Users

These are mainly associated with sample collection, sample pretreatments, and submission of an adequate amount of a sample to the ¹⁴C dating laboratory. The most important requirement of a user is to be able to define what constitutes a sample or an event to be dated and the association between the sample and the event. The likely age and possible contamination of a submitted sample as deduced from field information available would be of considerable advantage to the laboratory, as this will determine whether normal counting time needs to be lengthened or extensive pretreatment is required to remove contamination.

III. SOURCES OF ERROR, CORRECTION FACTORS, AND PRETREATMENTS

Errors associated with carbon dating can be conveniently divided into systematic errors and contamination errors.

A.   Systematic Errors and Correction Factors

These are due to the invalidity of the fundamental assumptions in the carbon dating method, such as temporal variations of the ¹⁴C production in the atmosphere (e.g., de Vries, 1958) and other causes (Lowe and Walker, 1984; Bradley, 1985).

Systematic errors are commonly overcome by using modern standards such as the international reference oxalic acid standard, SRM 4990C (U.S. National Bureau of Standards), or a reference value (e.g., some cockle shells, *Protothaca crassitesta*) (Rafter *et al.*, 1972). Corrections on secular effects based on dendrochronologically dated wood are applied using a variety of calibration curves (Fig. 2). and tables (e.g., Stuiver *et al.*, 1986).

Due to differences in the masses of carbon isotopes (¹⁴C, ¹³C, ¹²C) isotopic fractionation occurs during photosynthesis and in other physical, chemical, and biochemical processes. This is corrected by normalizing to a standard

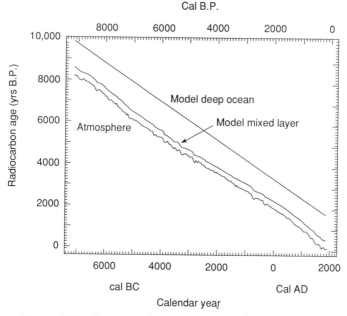

Fig. 2 Calibration curves for age conversion of samples. [Redrawn from Stuiver *et al.* (1986); by permission of *Radiocarbon.*]

$^{13}C/^{12}C$ ratio, which is the PDB limestone (belemnite carbonate from the Cretaceous Peedee Formation of South Carolina). Values of $\delta^{13}C$ are reported as deviations from this standard as:

$$\delta^{13}C\ (\text{‰}) = \frac{^{13}C/^{12}C\ \text{sample} - \ ^{13}C/^{12}C\ \text{standard (PDB)}}{^{13}C/^{12}C\ \text{standard (PDB)}} \times 1000.$$

Most biological materials tend to be depleted in heavy isotopes with $\delta^{13}C$ values ranging from -5‰ to -35‰ with the normal value of the standard being -25‰.

B. Contamination Errors and Pretreatments

In a contaminated sample, its $^{14}C/^{12}C$ ratio is not determined solely by radioactive decay since deposition but by the presence of foreign carbon, such as from plant roots and carbonates.

Errors associated with sample contamination are variable and difficult to overcome depending on the kind of materials and the locality where they were collected. In most instances, contamination may not be visible on inspection of the sample and drastic chemical pretreatments are necessary to remove contaminants. However, in some cases, the likelihood of contamination is evident by close examination of the site of collection (e.g., a wet site introduces foreign carbon by flowing water).

1. Causes and Kinds of Contamination

Several physical, chemical, and biological processes can affect contamination in biological materials used for dating under natural conditions (Stout *et al.*, 1981; Matthews, 1985). A common cause is adsorption of foreign organic carbon by soil-buried charcoals (Goh *et al.*, 1977a; Goh and Molloy, 1979). Humic acid contamination of soil-buried charcoals has also been reported by other workers (e.g., Bailey, 1971).

Exchanges of carbon between the sample and the atmosphere are a common cause of contamination in bone, wood, and shells. In some wooden specimens contamination can arise due to transportation of compounds across tree rings (Long *et al.*, 1979).

Materials (e.g., bone) buried anaerobically under natural conditions in the field (e.g., in a bog) when excavated and exposed to the air may exchange with atmospheric CO_2 and introduce young carbon into the sample. Contamination in shells can arise from the recrystallization of the shell carbonate in the field or laboratory due to the exchange of carbon with a different isotopic composition from that of the original shell carbon.

In the laboratory when a sample is treated with alkali in the presence of air, atmospheric CO_2 is absorbed by the alkali, thus trapping the carbon as contaminated carbonate. Organic solvents used in extracting lipids from soils and peats are often difficult to remove completely from the sample before ^{14}C dating. The adsorbed solvent carbon in the lipids often leads to young ^{14}C dates (Goh, 1978). Soil organic matter is a heterogeneous mixture of many components (Stevenson, 1982) showing a range of ^{14}C ages (Scharpenseel, 1971). Recent results showed that differences between ^{14}C ages of various humus fractions varied from soil to soil and with methods of fractionation (Stout *et al.*, 1981). It is therefore difficult to distinguish between contaminants and representative soil humic components. However, some successes have been achieved by treating soil organic matter with hot acids (e.g., $6M$ HCl or 70% HNO_3) and using the hydrolyzed residue for ^{14}C dating (Martel and Paul, 1974; Goh, 1978; Goh and Molloy, 1978).

Another kind of contamination is associated with the "hard-water" effect or "apparent age" (Bradley, 1985). This is due to aquatic plants or freshwater molluscs taking up carbon from water containing bicarbonate derived from old, inert sources (e.g., old peat bog or limestone and calcareous deposits) instead of carbon from the atmosphere. The problem is much more complicated with marine organisms as the turnover of oceanic water bodies is much slower than in the atmosphere (Baxter and Walter, 1971). The deep water remains out of contact with the atmosphere for centuries. Divergent surface currents may bring the deep water, depleted of ^{14}C (i.e., old carbon), into the surface causing an "apparent age" effect (Mangerud, 1972).

2. Elimination of Contamination

Methods of eliminating contamination are largely associated with the collection, storage, and preparation of samples before submitting for ^{14}C dating. These preventive measures involve mainly the selection of representative samples and proper packaging of samples.

Physical decontamination methods usually include the removal of visible roots and rootlets from soil samples by forceps under a low-power lens. Very fine rootlets may be removed by flotation while charcoal fragments are sometimes floated off from soil or sand. Dry and wet sieving are also used to separate charcoals or twigs into different particle sizes (Goh and Molloy, 1972).

Many ^{14}C dating laboratories conduct some forms of chemical treatments to remove carbonates such as treating samples with hot 2% phosphoric acid (Jansen, 1984) or acid wash with $1-2M$ HCl (Gillespie, 1984). More detailed chemical pretreatments are usually carried out by ^{14}C users and these are presented later.

IV. MATERIALS FOR CARBON DATING

A.   Suitability of Materials

In order to be suitable for use in carbon dating, materials commonly derived from living organisms must contain carbon that was initially able to exchange with the atmosphere. After the death of the organism, this carbon is not subjected to change other than radioactive decay. The material occurs in an environment conducive to its *in situ* preservation.

A wide range of materials can be used for ^{14}C dating. The most common materials are wood, charcoal, shell, bone, and peat. Other materials successfully dated include sediments from lakes and oceans, speleothems, and soil components, such as extractable and nonextractable soil organic matter fractions (e.g., humic acids, fulvic acids, nonextractable or nonhydrolyzable residues).

1. Wood

Traditionally, wood is usually considered to be an excellent dating material as it does not exchange readily with contemporary carbon. Simple decontamination procedures such as acid leach (hot HCl) to remove carbonate, or caustic leach (NaOH) to remove soil humic contaminants are assumed to be adequate for contaminated wood. For this reason, extensive decontamination procedures have not been investigated especially for wood buried in bogs for a long period.

It has been reported that the translocation of carbon compounds across tree rings can cause possible errors in dating woody materials (Long *et al.*, 1979). The recommendation is to isolate the holocellulose fraction for ^{14}C dating. A more recent procedure for badly contaminated wood is to chemically extract the cellulose fraction for dating (Gillespie, 1984). As the yield of cellulose from wood is usually low (<50%) a large amount of wood is required for extraction. For a large piece of well-preserved wood, it would be useful to separate it into different growth rings and date each separately.

2. Charcoal

Charcoals have long been considered as one of the most desirable materials for carbon dating because of their high carbon content, and the carbon when burnt does not suffer any mobilization or replacement by other carbon atoms (Geyh *et al.*, 1971; Keller and Rockwell, 1984). However, soil-buried charcoals can readily adsorb considerable amounts of soluble soil humic substances, formed either *in situ* or derived via pedotranslocation from associated soil horizons, especially in a bog or seasonally wet soil rich in organic matter (Goh and Molloy, 1979; Lee, 1981). Treating charcoals exhaustively with a mixture of $0.1M$ $Na_4P_2O_7$:$0.1M$ NaOH was found to be effective in removing contaminants (Goh and Molloy, 1972).

3. Peat

Peat is generally regarded as a reliable dating material (Geyh *et al.*, 1971) as good agreement has been obtained in ^{14}C dates of peat/wood and peat/charcoal pairs, and peat is normally formed *in situ*. However, erroneous ^{14}C dates between contemporary wood/peat pairs have been reported (Grant-Taylor, 1972). In order to overcome this problem, many laboratories have employed an alkaline wash to remove organic acids (Grant-Taylor and Rafter, 1972; Olsson and Florin, 1980). Other workers have dated the humic acid fraction after NaOH extraction (Polach and Singh, 1980). But in many of these studies, ^{14}C dates of humic acid fractions were not compared with the untreated peat (or control) or an insufficient range of extractants were used on the same material. When these factors were studied, it was found that considerable improvement in the reliability of dates was obtained after treatment with 70% HNO_3 (Goh, 1978; Hammond *et al.*, 1991).

4. Bone

Because of the porous nature of bone and its low carbon content, contamination is common, either from exchange of bone carbon with atmospheric modern carbon or ground water contamination enriched in ^{14}C (Olsson and

Broecker, 1958). Fossil bones are found to consist mainly of: (1) inorganic components (bone apatite and carbonate); and (2) organic components (collagen, >75%). The use of bone collagen (Longin, 1971; Gillespie, 1984) or amino acids, especially hydroxyproline (Gillespie et al., 1986), and not bone carbonate, is usually adopted as a decontamination measure.

5. Shell

Similar to bones, shells are very susceptible to environmental contamination. This usually arises from the exposure of shell inorganic carbonate to isotopic exchange and recrystallization of the metastable aragonite to stable calcite. If recrystallization occurs internally, it involves no contamination (Chappell and Polach, 1972). The routine procedure of decontamination is to leach the outer surface of the shell with dilute HCl or remove it mechanically by grinding followed by checking with x-ray diffraction or thin section microscopy to determine if calcite is present. In aragonite-secreting organisms, if calcite is detected, the specimens are rejected. However, for some marine organisms (e.g., oysters, barnacles, planktonic foraminifera) that secrete calcite into skeletons, their shells should give satisfactory dates. For these shells, their ^{14}C dates are corrected by using local shell standards (e.g., Prothothaca crassitesta).

6. Lake Sediments

Within lake bottoms, plant material originates from three main sources (Godwin, 1969): (1) dead phytoplankton and zooplankton (nektron mud); (2) detritus shed from plant communities around the lake fringes; and (3) material transported by streams and deposited as gel, mud (dy), or fine detritus land vegetation.

While materials from (2) and some of (3) derive their CO_2 from the air (thus, atmospheric radiocarbon activity), all the material from (1) and some of (3) may be sources of errors. This detritus may originate from the erosion of older peats and soils thereby introducing ancient material. Another source of error can arise when the lake is fed by water from ancient limestone or other sources of dissolved alkaline carbonate (e.g., lignites, shales, calcareous sedimentary rocks), causing the plankton to take up old dead carbon producing the hard water effect (Mathewes and Westgate, 1980).

Two other sources of errors can also arise. The relatively slow exchange of CO_2 between the lake reservoir and the atmosphere ("reservoir effect") results in the depletion of ^{14}C (Håkansson, 1979). Similar ^{14}C depletion can occur from effects of hydrothermal activity within a lake or in influent streams (Rafter et al., 1972).

Methods of removing contaminants in lake sediments before ^{14}C dating have not been studied extensively. Hence a recommended procedure does not exist other than a simple acid wash adopted by ^{14}C dating laboratories. Some workers recommend the use of lipids (Fowler *et al.*, 1986) and humic acids (e.g., Sheppard *et al.*, 1979).

7. Soils and Paleosols

These are usually dated as bulked samples without pretreatments. However, dates of bulked samples are usually considerably younger than those of pretreated samples (Goh and Molloy, 1978) and hence some form of pretreatment is necessary to yield even minimum dates.

The most reliable date is obtained by dating the soil residue after exhaustive hydrolysis with mineral acids (HCl, H_2SO_4, HNO_3) (e.g., Martel and Paul, 1974; Goh and Molloy, 1978; Hammond *et al.*, 1991). The soil used should be free of buried charcoals or dead carbon. The most appropriate acid to use depends on the carbon status of the residue after acid hydrolysis, as 70% HNO_3, being most drastic, may not leave sufficient carbon in the residue after hydrolysis for the conventional carbon dating technique (Goh and Pullar, 1977). This is not a problem using the new accelerator dating technique (Hammond *et al.*, 1991).

B. Sample Collection and Sample Size

The first decision to be made in the collection of materials for ^{14}C dating is to decide what constitutes a sample to date the event. The next step is to ensure that the sample collected is as representative as possible of the original sample which is most closely associated with the event. This is often made after careful examination of the original sample, the site and location where it is collected, and the relationship between the original sample and the event to be dated.

It is often advisable to take a photograph of the collection site for future reference and also to describe the location in as much detail as possible so that any worker can find the site readily. This information is later transferred to the sample submission form for recording and to inform the laboratory workers.

Another factor to be considered is the size of the sample to collect. This varies and is determined by the amount required by the carbon dating laboratory or its equipment and the carbon content of the sample.

Many carbon dating laboratories will process smaller samples than normally recommended but this decreases the precision and results in large errors or standard deviations. Minimum sample weights necessary for high precision dates are shown in Table 1. Conventional counting equipment

Table 1
Minimum Amount of Sample Required for Carbon Dating[a]

Sample material	Amount required	
	Conventional counters (g)	Accelerator counters (mg)
Wood (dry)	25–30	12–15
Wood (wet)	40–80	20–40
Charcoal (black)	12–20	6–10
Charcoal (brown)	50–100	25–50
Bone	50–100	25–50
Shell carbonate (hard)	90–100	45–50
Shell carbonate (powdery)	150–200	75–100
Peat, lake muds (dark brown, dry)	30–100	15–50
Peat, lake muds (light grey, dry)	120–200	60–100
Soil	2000–5000	1000–2500

[a] Data compiled from Jansen (1984) and Gillespie (1984).

requires larger samples (1–2 g C) than the new accelerator counters (5–10 mg C).

In practice, it is advisable to collect larger samples than required as this eliminates both the need to return to the collection site for more samples and the difficulties of relocating exactly the position of the collection area. Furthermore, the user can retain part of the sample to check anomalous results, to replace any likely loss of sample in the post or laboratory, and for other analyses if required.

After collection, samples should be handled and packaged carefully to avoid contamination. Ideal containers include strong polyethylene bags, aluminum foil, screw-top vials, or bottles of aluminum, polyethylene, or glass. These should be clean and dry. Wet samples (e.g., samples collected from sea water) should not be stored in plastic drums or wrappings, as PVA and PVC plastics contain plasticizers that can be absorbed by sample materials. Likewise, paper or cloth bags, thin sandwich bags, gelatin capsules, cotton wool, and paper tissues should not be used. When glass vials are used they should be packaged securely with padding to avoid breakage during transport to the laboratory. It is advisable to match sample size with container size to avoid the need of using packaging material. Glass wool is a suitable packaging material.

Samples should be correctly labeled with permanent-ink felt pens using file cards as labels rather than bits of paper or card. The label should contain a sample number and details of site, horizon and environment, and date of collection. Avoid using water-soluble ink felt pens and writing on the outside of plastic bags as these are likely to become illegible with time.

C. Sample Preparation

While out in the field, pick out and discard obvious foreign matter such as twigs, leaves, plant roots, stones, loose sand, or soil. If possible, physically separate marine from nonmarine shells, pick out and separate charcoal, bone, wood, and soil from each other.

In the laboratory, all these samples should be spread out on strong aluminum foil sheets and air dried. Visible plant roots and tissues are then removed under a low-power lens. Samples are then mixed thoroughly and subsampled if required. Soil, peat, and charcoal samples are usually crushed and sieved through a 2-mm sieve to remove roots and stones. Wood samples are dried and milled to sawdust (1 – 2 mm).

V. PRETREATMENT PROCEDURES AND CARBON DATING MEASUREMENTS

A. Pretreatment Procedures

The kind of pretreatments required is determined largely by the material to be dated as described previously.

1. Extraction of Wood Cellulose (Hesse, 1971)

Reagents:

Schweitzer's solution: Dissolve 10 g copper (II) sulphate in 100 ml of water and mix with a solution of 5 g KOH in 50 ml water. Collect the precipitate, wash with cold water, partially dry and macerate with 20 ml of a 20% ammonium hydroxide solution for 1 day with occasional shaking.

Ethanol 80%

Sodium hydroxide solution, 1% w/v, aqueous

Hydrogen peroxide, 20 volume

Hydrochloric acid, 30 ml of concentrated HCl per liter solution

Sodium hypochlorite solution

Procedure:

Heat 20 g of finely ground (1 – 2-mm sieve) wood or peaty material in a beaker with 200 ml of 1% NaOH solution. Boil the suspension for 30 min, maintaining the volume. Just acidify the mixture with HCl and allow the sample to settle; filter and wash the residue twice with hot distilled water. Replace the residue in the beaker and boil for 20 min with 200 ml of dilute HCl, maintaining the volume. Allow the residue to settle; filter and wash free of acid with hot distilled water. Replace the residue in the beaker, add 10 ml of sodium hypochlorite solution and bring the volume to 200 ml with distilled water. Allow to stand for 30 min with occasional shaking and maintaining the liquid alkaline (litmus paper test). Filter and treat the residue in a beaker with a further 10 ml of sodium hypochlorite and dilute to 200 ml with distilled water. After standing for 30 min, decant off the liquid and add 20 ml of H_2O_2. After settling, filter and wash the residue by decantation with hot distilled water; dry the washed residue on a steambath.

Crush the dry residue and mix it in a 300-ml centrifuge bottle with 200 ml of freshly prepared Schweitzer's solution. Shake for 6 hr and centifuge. Add 250 ml of 80% ethanol to 50 ml of extract and allow to stand overnight. Decant the liquid; filter and wash the precipitate (i.e., crude "cellulose") with hot water, 1M HCl, and water. Dry the precipitate of crude cellulose over a waterbath and pack it in a glass vial for dispatch to the carbon dating laboratory.

This extraction procedure is expected to lose more than half the weight of the original wood sample. If necessary, repeat the extraction on another lot of wood or peaty material to obtain sufficient cellulose for dating.

2. Extraction of Bone Collagen and Amino Acids (Longin, 1971; Gillespie *et al.,* 1986)

Reagents:

Hydrochloric acid: 0.6M, 2M, 6M solutions
Ammonia: 1.5M solution
Cation exchange resin (hydrogen form)

Wash the resin repeatedly (Amberlite IRC 50 H[+], 14–50 mesh or equivalent) with 2M NaOH and 2M HCl (Stevenson, 1965). Prepare a slurry of the resin in 2M HCl, and pour it into a 3.0 × 10-cm Pyrex glass tube with the lower end tapered to a small opening and provided with a filter made from a plug of Pyrex glass wool. Attach a piece of rubber tubing to the lower end of the tube. Wash the bed with distilled water until the effluent comes through neutral. Keep the surface of the resin wet as drying can bring about decomposition of the resin.

Procedure:

Clean surfaces of bone samples to remove encrustations by sandblasting with 30-μ alumina. Crush the sample in a stainless steel percussion mortar to pass through a 1-mm screen (e.g., a tea strainer). Place 100 g of the sample in a Pyrex culture tube with PTFE-lined cap and add 20 ml of 0.6M HCl. Shake or agitate the test tube with a vortex mixer occasionally for 2–24 hr. Centrifuge and discard the solution, repeat 2 or 3 times until no more gas is evolved and the solution pH remains below 3.0. Wash the sample with 20 ml distilled water and centrifuge again.

Transfer the sample to a round-bottom flask and add 5 ml of 6M HCl. Attach the flask to a reflux condenser, and boil the mixture under reflux (105°C) overnight (16–24 hr). To remove the color, dilute the solution to 20 ml with distilled water, add approximately 10 mg decolorizing charcoal, heat to 100°C for 10 min, cool, centrifuge, and retain the solution. To desalt the sample, pass the solution through a 10-ml column of cation-exchange resin, wash the column with 50 ml distilled water, discard the washing, then elute with 20 ml of 1.5M ammonia solution. Collect the ammonia solution in a PTFE beaker; evaporate under a heat pump with purified air flow. Transfer the last 0.5–1 ml to a combustion tube for final drying at 80°C with air flow. Convert the sample to graphite for AMS measurement as described by Gillespie *et al.* (1984).

3. Extraction of Shell Carbonate or Organic Fractions (Gillespie *et al.,* 1986)

Reagents:

    Phosphoric acid: 50% solution

    Other reagents as described for bone collagen and amino acids above.

Procedure:

Clean the shell fragments by sandblasting and etch to remove 10 to 20% of surface carbonate with dilute HCl as described for bone samples. Treat the cleaned shell fragments with 50% phosphoric acid under vacuum. Collect the carbon dioxide released and convert it to graphite for AMS counting as described by Gillespie *et al.* (1984).

To extract the organic fractions for dating, treat 15-g shell samples as follows:

1. Add excess 2M HCl to the shell sample to dissolve the carbonate matrix. Wash the insoluble residue with distilled water.

2. Hydrolyze the residue in 6M HCl at 105°C overnight as described above for bone samples. Recover the insoluble dark residue or "humic acid" by filtering. Dry the residue and submit for dating.

3. Desalt the soluble hydrolysate using cation-exchange resin; dry and submit for dating as described for bone amino acids.

4. Pretreatment for Charcoals (Goh and Molloy, 1972)

Reagents:

    Sodium pyrophosphate: sodium hydroxide solution [sodium pyrophosphate: $0.1M$ $Na_4P_2O_7$ (44.6 g $Na_4P_2O_7$/l); sodium hydroxide: $0.1M$ NaOH (4 g NaOH/l) mixture]

    Hydrochloric acid; 3M solution

Procedure:

Add 80 ml of $0.1M$ $Na_4P_2O_7$: $0.1M$ NaOH solution to 200 g of finely ground (60-mesh or 250-μm sieve) air-dried charcoals in a polythene centrifuge bottle or beaker. Stir the suspension for about 5–10 min, then allow it to settle overnight in an atmosphere of nitrogen (pass nitrogen gas into the beaker and then close the lid). The next day, centrifuge the suspension and recover the residue. Repeat the above extraction on the residue until the extract obtained is not colored. Treat the residue 3 times with 3M HCl followed once by distilled water. Dry the residue, grind it to pass through a 60-mesh (250-μm) sieve before sending it to the carbon dating laboratory.

5. Lipid Extraction (Goh, 1978).

Reagents:
   Hydrochloric acid : hydrofluoric acid (2.5% HCl : 2.5% HF)
   Acetone
   Petroleum ether (40° – 60°C)
   Ethanol : benzene (1 : 4)
   Methanol : chloroform (1 : 1)
   Chloroform
   Ether : petroleum (1 : 1)

Procedure:

Add 10 ml of HCl : HF solution to 1 g of air-dried finely ground (60-mesh or 250-μm sieve) soil or peat in a polyethylene centrifuge tube; stopper and shake overnight. Centrifuge the suspension and discard the supernatant liquid. Repeat the above treatment. Wash the soil residue with 10-ml portions of distilled water until acid free and discard the washings. Add 50 ml of acetone to the soil residue, allow to stand for 4 hr while covered and stir every 20–30 min. Centrifuge and transfer the solution to a 500-ml bottle fitted with a ground glass stopper. Repeat the above extraction using 50 ml of petroleum, 50 ml of ethanol : benzene, and 50 ml methanol : chloroform.

Bulk all these extracts and evaporate slowly at low temperature and pressure to dryness. Reextract the resultant residue with 20 ml of cold ether : chloroform for 2 min; 20 ml cold chloroform for 2 min; 20 ml ether : petroleum; bring to a boil on a waterbath, then decant; add 20 ml chloroform, bring to a boil, and then decant. Bulk all these extracts and evaporate to dryness on a waterbath in a 250-ml silica beaker. Grind the residue to pass through a 60-mesh (250-μm) sieve before sending to the carbon dating laboratory. If insufficient lipids are extracted, use a larger amount of soil or peat for extraction and increase the amounts of extractants proportionally.

6. Extractions of Soil Components

a. Nitric Acid Hydrolysis

Reagents:
   Nitric acid: 70% solution and 6M solution

Procedure:

Add 1 l of 70% HNO_3 to 40 g of air-dried, finely ground (<2-mm sieve) soil (1:25 soil:acid ratio) in a beaker (2-l capacity). Cover the beaker and heat on a hot plate in the fume hood or cupboard. Allow the solution to boil for 20 min, taking care to avoid overboiling or sputtering. Cool and add 185 ml of $6M$ HNO_3. Allow the mixture to stand overnight while covered.

Filter the solution through a sintered glass Buchner funnel. Discard the filtrate. Wash the residue with distilled water and then dry it under vacuum. Grind the residue to pass through a 60-mesh (250-μm) sieve before sending to the carbon dating laboratory.

b. Humic Acid Extraction (Goh and Molloy, 1978)

Reagents:

Sodium pyrophospate:sodium hydroxide (0.1M $Na_4P_2O_7$:0.1M NaOH)

Hydrochloric acid:6M and 3M solutions

Procedure:

Add 400 ml of 0.1M $Na_4P_2O_7$:0.1M NaOH to 10 g of air-dried and finely ground (<2-mm sieve) soil in a polythene bottle. Shake the suspension for 5–10 min by hand or stir thoroughly. Allow the suspension to settle overnight in an atmosphere of nitrogen gas by passing nitrogen gas into the bottle to displace air. The next day, centrifuge the suspension and recover the supernatant. Neutralize the supernatant to pH 7.0 with 3M HCl and store it in a polythene bottle in a nitrogen gas atmosphere.

Repeat the extraction (3–40 times) with the residue until the extract is no longer colored. Combine all the extracts and add 6M HCl gradually to bring the pH to 1.0. Allow the precipitate formed, which is humic acid, to settle overnight. Centrifuge the suspension and recover the humic acid. Wash the humic acid 3 times with 3M HCl followed once with distilled water. Dry the humic acid, grind, and send to the carbon dating laboratory.

B. Carbon Dating Measurements

1. Conventional Carbon Dating Techniques

a. Gas Proportional Counting Technique (Rafter, 1965)

In this method, carbon in the sample is converted into a gas (carbon dioxide, methane, or acetylene), which is then introduced into a proportional counter (counter output voltage pulses are proportional to β radiation received) to measure β-particle emissions. To convert to carbon dioxide, the sample is combusted in an excess of oxygen in an inconel tube furnace packed with an oxidizing agent (e.g., CuO) at 780°C. The combustion gases are then passed through a series of wash bottles to remove impurities and water. Vapors of CO_2 are then trapped in liquid nitrogen and subjected to further purification and drying with dry ice/alcohol traps before being stored in copper storage tanks and held for at least 24 hr to allow for the decay of any trace radon contamination. The CO_2 purity is checked and repurified if necessary. From the storage cylinder, CO_2 is condensed back into a cold finger by liquid nitrogen and then expanded through P_2O_5 traps into a second cold finger before being expanded into the counter for counting. The sample is counted for a minimum time of 1000–3000 min.

Detailed procedures for converting the carbon of a sample to methane and benzene instead of CO_2 are available elsewhere (e.g., Polach and Stipp, 1967).

b. Liquid Scintillation Counting Technique (Hogg, 1982)   Organic carbon in a sample is combusted in a fused silica combustion tube in the presence of excess oxygen and CuO. The CO_2 produced is purified by passage through the purification train and then collected in liquid N_2 traps. Inorganic carbon is hydrolyzed by $2N$ HCl and the CO_2 is collected after the removal of water in ethanol-liquid N_2 traps and a silica gel column.

The CO_2 is then initially converted to lithium carbide, which is hydrolyzed to acetylene followed by purification and drying before being collected in liquid N_2 cold traps. The frozen acetylene is allowed to sublime and subjected to drying by passage through an ethanol-liquid N_2 trap. It is then passed through a vanadium-activated silica–alumina catalyst column (Noakes *et al.*, 1965) to trimerize the acetylene to benzene. The catalyst column is heated to 100°C and benzene is collected under vacuum in an isopropyl dry ice trap, then transferred to a sample vial to await counting.

Benzene samples are transferred to low-^{40}K silica vials (5-ml capacity) and mixed with scintillator reagents. The vials are sealed with teflon-lined stoppers and the samples are counted in a liquid scintillation counter.

2. Accelerator Mass Spectrometric Technique (AMS) (Wallace *et al.,* 1987)

This technique differs from the conventional carbon dating techniques as it does not count radioactive disintegrations of ^{14}C atoms but the relative number of ^{14}C, ^{13}C, and ^{12}C atoms contained in solid carbon samples (e.g., graphite target) using an ultrasensitive mass spectrometer. A full-size particle accelerator (e.g., tandem accelerator) is used to impart high energies to the carbon ions. This technique enables positive identification of the carbon ions and discriminates against other ions of similar masses.

Samples are converted to CO_2 by the standard combustion technique as described above. The CO_2 is then converted to graphite pellets using excess H_2 and an iron catalyst (Lowe and Judd, 1987). The graphite is packed into a copper rod well. It is then exposed to a Cs-sputter ion beam and the different ions of carbon are discriminated and measured by the mass spectrometer.

VI. CALCULATING, REPORTING, AND INTERPRETING ^{14}C AGE

^{14}C ages are calculated as the elapsed time t using the expression

$$t = Tm \ln \frac{A_o}{A_s}$$

where t = age in years, Tm = mean life of ^{14}C, A_o = specific activity of oxalic acid standard, SRM 4990C, and A_s = specific activity of sample.

Different carbon dating laboratories adopt different practices in reporting their results. These are published in the *Radiocarbon* journal and should be consulted if ^{14}C ages from different laboratories are to be compared.

In general, ^{14}C results are reported as:

age ± error in years before present (B.P.).

The age is calculated usually using the old ($t_{1/2} = 5568 \pm 30$) instead of the new half-life ($t_{1/2} = 5730 \pm 40$). It is expressed relative to the date of the 0.95 oxalic acid standard in the form of years B.P. (i.e., years before 1950). Ages are rounded off as shown in Table 2.

Errors are usually reported as one standard deviation, unless otherwise stated. For samples with ages greater than 35,000 years, positive and negative errors are shown separately because of the asymmetry of the error. For ages

Table 2
Rounding of Carbon-14 Ages[a]

Age-range (years)	Rounded to nearest years	Standard deviation rounded up to nearest years	Precision (%)
1–1000	1	1	>0.1
1000–2000	5	5	0.25–0.50
2000–10,000	10	10	0.10–0.50
10,000–20,000	50	50	0.25–0.50
20,000 and up	100	100	>0.50

[a] From Jansen (1984).

of less than 200 years B.P., the percentage modern carbon is shown and is estimated by the expression

$$\text{modern C (\%)} = 100 \exp\left(\frac{-\text{age}}{Tm}\right).$$

The error as quoted is derived from a consideration of counting statistics and determination of background, sample, and standards. This error also depends on sample activity as a young sample with high activity has a smaller error than an old sample with low activity. The quoted error does not refer to the sampling or field error, which is the error obtained in dating a series of repeats of independent preparations of a sample or samples from within a site. This is an important error and should be determined in routine carbon dating.

Carbon-14 ages are often reported after correction for secular effects to correct for systematic errors as discussed earlier. The conversion factors for converting ages from "new" to "old" mean life and vice versa are:

age to new mean life = age to old mean life × 1.029095, and

age to old mean life = age to new mean life × 0.9717277.

It is important to bear in mind that ^{14}C years are not the same as calendar years because of difficulties in establishing accurately the half-life of ^{14}C and also the invalidity of the basic asumptions of the ^{14}C method of a constant rate of ^{14}C production and other secular effects as mentioned earlier.

Because soil organic matter consists of multi-components of humic substances of different ages (Stevenson, 1982) and is constantly being added with time, especially on surface soils, a number of concepts have been used to express the ^{14}C ages of soils (Stout et $al.$, 1981). According to Geyh et $al.$ (1971), the mean age of a soil is the arithmetic average of two ages, one at the beginning (A_1) and the other end of a soil-forming process:

$$A = \frac{A_1 + A_2}{2}$$

The mean residence time (MRT) $\bar{a}$ is defined as the weighted mean of the different ages (A) of the different soil organic matter fractions present in different amounts as:

$$\bar{a} = \frac{\int_1^2 f(A)A\,dA}{\int_1^2 f(A)\,dA}.$$

The mean true residence time of a soil is calculated as:

$$\bar{a} = \frac{2c(A_1^3 - A_2^3) - 3B(A_1^2 - A_2^2)}{3c(A_1^2 - A_2^2) - 6B(A_1 - A_2)} + A_2$$

where

$$f(A) = B - cA,$$

B = quantity of the fraction with age 0, and

c = increase in the amount of organic matter per unit time.

A linear relationship is assumed for changes in the amounts of the various organic matter fractions. More recent studies have shown this is not strictly true.

Perrin *et al.* (1964) have postulated that the ^{14}C age of a soil is an integration of accumulation over a long time period and is not likely to represent the real age. Three factors affect the apparent age and these are shown in the expression:

$$X = \int {}^t_0 I_0 \alpha c^{-\lambda} dt$$

where

I_0 = present ^{14}C activity,

λ = decay constant, and

α = rate of eluviation.

It is evident that it is difficult to interpret soil ages because of the possible rejuvenation of soil organic matter and the continual addition of fresh plant material. Ages obtained from dating residue of soil organic matter after acid hydrolysis (Goh and Molloy, 1978) are generally regarded as minimum ages.

REFERENCES

Bailey, J. M. (1971). Extractions and radiocarbon dating of dispersed organic materials from loess in the South Island of New Zealand. *New Zealand J. Sci.* **14,** 490–493.

Baxter, M. S., and Walter, A. (1971). Fluctuations of atmospheric carbon-14 concentrations during the past century. *Proc. Royal Soc. London,* Ser. A, **321,** 105–127.

Bradley, R. D. (1985). "Quaternary Paleoclimate: Methods of Paleoclimate Reconstruction." George Allen and Unwin, Boston.

Chappell, J. M. A., and Polach, H. A. (1972). Some effects of partial recrystallization on ^{14}C dating of late Pleistocene corals and molluscs. *Quat. Res.* **2,** 244–252.

de Vries, H. (1958). Variation in concentration of radiocarbon with time and location on earth. *Koninkl Nederlandse Wetensch Proc.* Ser. B, **61,** 94–102.

Fowler, A. J., Gillespie, R., and Hedges, R. E. M. (1986). Radiocarbon dating of sediments. *Radiocarbon* **28,** 441–450.

Geyh, M. A., Benzler, J. H., and Roeschmann, G. (1971). Problems of dating Pleistocene and Holocene soils by radiometric methods. In "Paleopedology" (D. H. Yaalon, ed.), pp. 63–75. Israel Univ. Press, Jerusalem.

Gillespie, R. (1984). "Radiocarbon Users' Handbook." Oxford Univ. Committee for Archaeol., Oxbow Books, Oxford.

Gillespie, R., Hedges, R. E. M., and Wand, J. O. (1984). Radiocarbon dating of bone by accelerator mass spectrometry. *J. Archaol. Sci.* **11,** 165–170.

Gillespie, R., Hedges, R. E. M., and Humm, M. J. (1986). Routine AMS dating of bone and shell proteins. *Radiocarbon* **28**, 451–456.

Godwin, H. (1969). The value of plant materials for radiocarbon dating. *Am. J. Bot.* **56**, 723–731.

Goh, K. M. (1978). Removal of contaminants to improve the reliability of radiocarbon dates of peats. *J. Soil Sci.* **29**, 340–349.

Goh, K. M., and Molloy, B. P. J. (1972). Reliability of radiocarbon dates from buried charcoals. *Proc. 8th Intern. Conf. on Radiocarbon Dating*, pp. 565–581. Roy. Soc. New Zealand, Wellington.

Goh, K. M., and Molloy, B. P. J. (1978). Radiocarbon dating of paleosols using soil organic matter components. *J. Soil Sci.* **29**, 567–573.

Goh, K. M., and Molloy, B. P. J. (1979). Contamination in charcoals used for radiocarbon dating. *New Zealand J. Sci.* **22**, 39–47.

Goh, K. M., and Pullar, W. A. (1977). Radiocarbon dating techniques for tephras in central North Island, New Zealand. *Geoderma* **18**, 263–278.

Goh, K. M., Molloy, B. P. J., and Rafter, T. A. (1977a). Radiocarbon dating of Quaternary loess deposits, Banks Peninsula, Canterbury, New Zealand. *Quat. Res.* **7**, 177–196.

Goh, K. M., Stout, J. D., and Rafter, T. A. (1977b). Radiocarbon enrichment of soil organic matter fractions in New Zealand soils. *Soil Sci.* **123**, 385–391.

Grant-Taylor, T. L. (1972). The extraction and use of plant lipids as a material for radiocarbon dating. *Proc. 8th Intern. Conf. on Radiocarbon Dating*, pp. 439–449. Roy. Soc. New Zealand, Wellington.

Grant-Taylor, T. L., and Rafter, T. A. (1972). New Zealand radiocarbon age measurements. *New Zealand J. Geol. and Geophys.* **14**, 364–402.

Håkansson, S. (1979). Radiocarbon activity in submerged plants from various South Swedish lakes. *Proc. 9th Intern. Radiocarbon Conf.*, pp. 433–443. Univ. Calif. Press, Berkeley.

Hammond, A. P., Goh, K. M., Tonkin, P. J., and Manning, M. R. (1991). Chemical pretreatments for improving the radiocarbon dates of peats and organic silts and a gley Podzol environment: Graham's terrace, north Westland, New Zealand. *New Zealand J. Geol. and Geophys.* (in press). Nelson.

Hesse, P. R. (1971). "A Textbook of Soil Chemical Analysis." John Murray Pub., London.

Hogg, A. G. (1982). Radiocarbon dating at the University of Waikato, New Zealand. Occasion Report Number 8. University of Waikato, Department of Earth Sciences.

Jansen, H. S. (1984). "Radiocarbon Dating for Contributors." Institute of Nuclear Sciences, Report No. 328, D.S.I.R., Lower Hutt, New Zealand.

Keller, E. A., and Rockwell, T. K. (1984). Tectonic geomorphology, Quaternary chronology and paleoseismicity. In "Developments and Application of Geomorphology." (J. E. Costa and P. J. Fleisher, eds.), pp. 203–339. Springer-Verlag, Berlin.

Lee, R. E. (1981). Radiocarbon ages in error. *Anthrop J. Canada* **19**, 9–29.

Long, A., Arnold, L. D., Damon, P. E., Ferguson, C. W., Lerman, J. C., and Wilson, T. A. (1979). Radial translocation of carbon in bristlecone pine. *Proc. 9th Intern. Radiocarbon Conf.*, pp. 532–537. Univ. Calif. Press, Berkeley.

Longin, R. (1971). New method of collagen extraction for radiocarbon dating. *Nature (London)* **230**, 241–242.

Lowe, D. C., and Judd, W. J. (1987). Graphite target preparation for radiocarbon dating by accelerator mass spectrometry. *Nuclear Instruments and Methods in Phys. Res.* **B28**, 113–116.

Lowe, J. J., and Walker, M. J. C. (1984). "Reconstructing Quaternary Environments." Longman, London.

Mangerud, J. (1972). Radiocarbon dating of marine shells including a discussion of apparent age of recent shells from Norway. *Boreas* 1, 143–172.

Martel, Y. A., and Paul, E. A. (1974). The use of radiocarbon dating of organic matter in the study of soil genesis. *Soil Sci. Soc. Am. Proc.* 38, 501–506.

Mathewes, R. W., and Westgate, J. A. (1980). Bridge River tephra: revised distribution and significance for detecting old carbon errors in radiocarbon dates of limnic sediments in southern British Columbia. *Canad. J. Earth Sci.* 17, 1454–1461.

Matthews, J. A. (1985). Radiocarbon dating of surface and buried soils: principles, problems and prospects. In "Geomorphology and Soils." (R. S. Richards, R. R. Arnett, and S. Ellis, eds.), pp. 269–288. George Allen and Unwin, London.

Noakes, J. E., Kim, S. M., and Stipp, J. J. (1965). Chemical and counting advances in liquid scintillation age dating. *Proc. 6th Intern. Conf. on Radiocarbon and Tritium Dating*, pp. 68–92. Clearinghouse for Fed. Sci. of Tech. Inf. Natl. Bur. Standards, Washington, D.C.

Olsson, E. A., and Broecker, W. S. (1958). Sample contamination and reliability of radiocarbon dates. *New York Acad. Sci. Trans.* 20, 593–604.

Olsson, I. U., and Florin, M. B. (1980). Radiocarbon dating of clay and peat in the Getsjö area, Kolmården, Sweden, to determine rational limit of *Picea*. *Boreas* 9, 289–305.

Perrin, R. M. S., Willis, E. H., and Hodge, C. A. H. (1964). Dating of humus Podzols by residual radiocarbon activity. *Nature (London)* 202, 165–166.

Polach, H. A., and Singh, G. (1980). Contemporary ^{14}C levels and their significance to sedimentary history of Bega Swamp, New South Wales. *Radiocarbon* 22, 398–409.

Polach, H. A., and Stipp, J. J. (1967). Improved synthesis techniques for methane and benzene radiocarbon dating. *Intern. J. Appl. Rad. Isotop.* 18, 359–364.

Rafter, T. A. (1965). Carbon-14 variation in nature. Part 1. Technique of ^{14}C preparation, counting, and reporting of results. *New Zealand J. Sci.* 18, 359–364.

Rafter, T. A., Jansen, H. S., Lockeridge, L., and Trotter, M. M. (1972). New Zealand radiocarbon reference standards. *Proc. 8th Intern. Conf. on Radiocarbon Dating*, pp. 625–675. Roy. Soc. New Zealand, Wellington.

Scharpenseel, H. W. (1971). Radiocarbon dating of soils, problems, troubles, hopes. In "Paleopedology" (D. H. Yaalon, ed.), pp. 77–88. Israel Univ. Press, Jerusalem.

Sheppard, J. C., Ali, S. Y., and Mehringer, Jr., P. J. (1979). Radiocarbon dating of organic components of sediments and peats. *Proc. 9th Intern. Radiocarbon Conf.*, pp. 284–305. University of California Press, Berkeley.

Stevenson, F. J. (1965). Amino acids. In "Methods of Soil Analysis," Part 2 (C. A. Black, ed.), pp. 1437–1450. Am. Soc. Agron., Madison, WI.

Stevenson, F. J. (1982). "Humus Chemistry: Genesis, Composition, Reactions." Wiley-Intersci., New York.

Stout, J. D., Goh, K. M., and Rafter, T. A. (1981). Chemistry and turnover of naturally occurring resistant organic compounds in soil. *Soil Biochem.* 5, 1–73.

Stuiver, M., Pearson, G. W., and Braziunas, T. (1986). Radiocarbon age calibration of marine samples back to 9000 Cal. Yr. B.P. *Radiocarbon* 28, 980–1021.

Wallace, G., Sparks, R. J., Lowe, D. C., and Pohl, K. P. (1987). The New Zealand accelerator mass spectrometry facility. *Nuclear Instruments and Methods in Phys. Res.* B29, 124–128.

9

Bomb Carbon

K. M. Goh

Department of Soil Science
Lincoln University
Canterbury, New Zealand

I. INTRODUCTION

The detonation of thermonuclear devices in the 1950s and 1960s has resulted in the enrichment of the atmosphere with bomb ^{14}C. Maximum increases in atmospheric ^{14}C levels occurred in 1961 and 1962, accounting for more than 70% of the total (Lassey *et al.*, 1987). At present, the enrichment has ceased and is depleting at the rate of about 6.1% per year.

Many recent studies have shown that bomb ^{14}C represents a spike input of ^{14}C to the stratosphere, which subsequently entered the terrestrial ecosystems through plants and then recycled through animals, microorganisms, soils, and soil organic matter (e.g., Rafter and Stout, 1970; Scharpenseel, 1973; Goh *et al.*, 1977; O'Brien and Stout, 1978; Stout *et al.*, 1981; Anderson and Paul, 1984; Goh *et al.*, 1984). This bomb carbon can be used to study the dynamics of soil organic matter. By measuring the bomb ^{14}C content of soil organic carbon present in the soil profile, and assuming steady state conditions exist, a number of organic carbon dynamics such as annual input of carbon (F), rate of decomposition (λ), turnover time, (τ, which is the inverse of λ), and downward movement or diffusivity (κ) of carbon in the soil profile have been determined (O'Brien and Stout, 1978; O'Brien, 1984; 1986).

II. GENERAL AND SPECIFIC REQUIREMENTS

Requirements of ^{14}C laboratories and users as described for carbon dating apply also to bomb ^{14}C enrichment studies (see Chapter 8, this volume). However, in addition, specific requirements are needed in the bomb ^{14}C enrichment technique. These are largely related to the requirement of determining soil density and the assumptions of the model used to calculate the various dynamic parameters.

The most widely used model is a steady state diffusion model developed by O'Brien and Stout (1978). This model assumes that a fixed and steady rate applies to the continual entry of fresh carbon into the soil surface, the transport of carbon by diffusion downwards through the soil profile, and the loss of carbon by respiration (O'Brien, 1984). These assumptions need to be verified by ^{14}C data from a soil profile that has not been exposed to bomb ^{14}C enrichment. The unexposed soil profile should be of the same soil type and should preferably occur adjacent to the site of interest (i.e., exposed profile). It is usually obtained from a soil beneath a structure (e.g. house) built before 1950 (O'Brien, 1986) or from the deepest horizon of an exposed soil profile which has not been contaminated with organic carbon enriched with ^{14}C moving from soil horizons above (Shoenau and Bettany, 1987).

The carbon content of each soil horizon is expressed as a percentage and the soil bulk density needs to be measured or assumed to be unity. The total soil ^{14}C activity is treated as a function of:

1. "old" decayed carbon with a ^{14}C activity less than that of the 0.95 NBS oxalic acid standard,
2. "modern" carbon with a ^{14}C activity significantly greater than the 0.95 oxalic acid standard (i.e., 100% modern), and
3. "bomb" carbon with a ^{14}C activity significantly greater than the 0.95 oxalic acid standard.

To study the effects of earthworms on soil organic matter dynamics, ^{14}C data of soils in the presence and absence of earthworms are compared (Stout and Goh, 1980). The incorporation of bomb ^{14}C into the various soil organic matter fractions (e.g., fulvic acid, humic acid, humin, acid hydrolysate, and acid residue) (Goh et al., 1977; Goh et al., 1984) can be obtained by preparing these fractions, using the same methods as described for carbon dating.

III. SUITABILITY OF MATERIALS

All biological materials (e.g., soils, plants, earthworms) suitable for carbon dating are also suitable for bomb carbon enrichment studies.

IV. FIELD AND LABORATORY PROCEDURES

The field and laboratory procedures as described for the collection and treatments of biological materials for carbon dating are used for bomb ^{14}C enrichment studies.

With soils, soil cores of known volume can be sampled using manual or mechanical corers in the field. The bulk density is determined on soils dried to a constant weight at 105°C and is calculated as mass of dry soil per unit volume, usually gcm^{-3}.

The annual input of carbon (F) is obtained by measuring the annual amount of plant residues returning to the soil. The root growth is measured or assumed to be a definite proportion of plant top growth depending on plant species and growth conditions. The carbon contents of plant tissues and soils are determined by standard combustion technique (Goh and Heng, 1987).

Measurements of ^{14}C activity of samples are conducted using the same techniques (e.g., gas proportional counting, accelerator mass spectrometry) as described for carbon dating.

V. COMPUTATIONS OF BOMB ^{14}C RESULTS

A. Parameters

The following parameters of soil organic matter dynamics can be computed as:

1. Total amount of bomb ^{14}C in a soil profile (T_k)

$$T_k = \sum_j l_j P_j C_j \Delta^{14}C_j'$$

where

$l_j =$ depth of horizon j in cm,
$P_j =$ soil bulk density in horizon j,
$C_j =$ per cent organic carbon in horizon j,
$\Delta^{14}C' =$ increase in ^{14}C in horizon j due to incorporation of bomb ^{14}C,
$\Sigma =$ summation of all horizons that show a significant increase in ^{14}C, and
$k =$ year in which the soil was sampled.

2. Total amount of organic carbon per unit area in the soil profile (W)

$$W = \sum_j l_j P_j C_j$$

3. Turnover time (τ)

$$\tau = W/F$$

4. Rate of decomposition or respiration (λ)

$$\lambda = \frac{1}{\tau}$$

5. Mean depth of soil carbon (Z_0)

$$Z_0 = 100 \ W/C_0P_0$$

where
   C_0 = per cent carbon in the topsoil horizon, and
   P_0 = soil bulk density in the topsoil horizon.
6. Carbon diffusivity (κ) in the soil profile

$$\kappa = Z_0^2/\tau$$

B. Verification of Assumptions in the Steady State Diffusion Model

Assumptions of the steady state model can be verified only when ^{14}C data are available from a soil profile or soil horizon that has not been exposed to bomb ^{14}C and which is adjacent to or of the same soil type as the soil profile that has been exposed to bomb ^{14}C. Alternatively, an exposed soil profile with a known rate of ^{14}C enrichment with respect to time can be used in place of the unexposed soil.

1. Verification of the assumption of constant amount of "old" carbon with soil depth. Estimate the "old" carbon as:

   "old" carbon (%) = [carbon (%) $\times \Delta^{14}C$]/100.

   Estimate this for each horizon for both exposed and unexposed soil profiles and observe for constancy.
2. Estimation of increase in $\Delta^{14}C$ due to atom bomb ^{14}C input in exposed soil profile (i.e., $\Delta^{14}C''$) is

$$\Delta^{14}C'' = \Delta^{14}C - \Delta^{14}C'$$

where
   $\Delta^{14}C = \Delta^{14}C$ in exposed soil horizon, and
   $\Delta^{14}C' = [\Delta^{14}C^* \times carbon^* \ (\%)]/carbon^{**} \ (\%)$.

* From unexposed horizon; ** from corresponding exposed horizon.

REFERENCES

Anderson, D. W., and Paul, E. A. (1984). Organo-mineral complexes and their study by radiocarbon dating. *Soil Sci. Soc. Am. J.* 84, 298–301.

Goh, K. M., and Heng, S. (1987). The quantity and nature of the forest floor and topsoil under some indigenous forests and nearby areas converted to *Pinus radiata* plantations in South Island, New Zealand. *New Zealand J. Botany* 25, 243–254.

Goh, K. M., Stout, J. D., and Rafter, T. A. (1977). Radiocarbon enrichment of soil organic matter fractions in New Zealand soils. *Soil Sci.* 123, 385–391.

Goh, K. M., Stout, J. D., and O'Brien, B. J. (1984). The significance of fractionation in dating the age and turnover of soil organic matter. *New Zealand J. Sci.* 27, 69–72.

Lassey, K. R., Manning, M. R., and O'Brien, B. J. (1987). An assessment of the inventory of carbon-14 in the oceans: an overview. Institute of Nuclear Sciences INS-R-368, D.S.I.R., Lower Hutt, New Zealand.

O'Brien, B. J. (1984). Soil organic carbon fluxes and turnover rates estimated from radiocarbon enrichments. *Soil Biol. Biochem.* 16, 115–120.

O'Brien, B. J. (1986). The use of natural and anthropogenic ^{14}C to investigate the dynamics of soil organic carbon. *Radiocarbon* 28, 358–362.

O'Brien, B. J., and Stout, J. D. (1978). Movement and turnover of soil organic matter as indicated by carbon isotope measurements. *Soil Biol. Biochem.* 10, 309–317.

Rafter, T. A., and Stout, J. D. (1970). Radiocarbon measurements as an index of the rate of turnover of organic matter in forest and grassland ecosystems in New Zealand. In "Radiocarbon Variations and Absolute Chronology." (I. U. Olsson, ed.), pp. 401–418. Nobel Symposium, 7th Proc., Stockholm, Sweden.

Scharpenseel, H. W. (1973). Natural radiocarbon measurements on soil organic matter fractions and on soil profiles of different pedogenesis. *Proc. 8th Intern. Radiocarbon Dating Conf.* 2, 383–393. Royal Soc. New Zealand, Wellington.

Shoenau, J. J., and Bettany, J. R. (1987). Organic matter leaching as a component of carbon, nitrogen, phosphorus, and sulfur cycles in a forest, grassland and gleyed soil. *Soil Sci. Soc. Am. J.* 51, 646–651.

Stout, J. D., and Goh, K. M. (1980). The use of radiocarbon to measure the effects of earthworms on soil development. *Radiocarbon* 22, 892–896.

Stout, J. D., Goh, K. M., and Rafter, T. A. (1981). Chemistry and turnover of naturally occurring resistant organic compounds in soil. *Soil Biochem.* 5, 1–73.

II

Uses and Procedures for ^{13}C

10

Stable Carbon Isotope Ratios of Natural Materials: I. Sample Preparation and Mass Spectrometric Analysis

Thomas W. Boutton

Department of Rangeland Ecology and Management
Texas Agricultural Experiment Station
Texas A&M University
College Station, Texas 77843

I. INTRODUCTION

Approximately 98.89% of all carbon in nature is ^{12}C, and 1.11% of all carbon is ^{13}C. The ratio of these two stable isotopes in natural materials varies slightly around these average values as a result of isotopic fractionation during physical, chemical, and biological processes. In materials of biological interest, this variation is relatively small, with the most enriched materials (those highest in ^{13}C) differing from the least enriched (those lowest in ^{13}C) by only about 10% or 100 parts per thousand. However, differences between materials in the range of 1 to 10 parts per thousand in ^{13}C content are often significant to biologists, and differences of 0.1 parts per thousand are often significant to paleoclimatologists and geochemists.

II. ISOTOPE RATIO MASS SPECTROMETRY

In order to make use of these small but significant variations, stable carbon isotope ratios ($^{13}C/^{12}C$) must be measured with extremely high precision. The only type of instrument currently capable of such high precision measurements is a dual-inlet gas isotope ratio mass spectrometer (GIRMS) equipped with two or more ion beam collectors. This type of instrument was first described in 1950 by McKinney et al., but modern state-of-the-art instruments are of the same basic design. Since 1950, advances in electronics, ion optics, and vacuum technology have improved the attainable precision by approximately tenfold. In addition, most commercially available instruments are now fully automated and permit unattended analyses of 10 to 80 samples per day.

The high precision afforded by these mass spectrometers is due to simultaneous collection of the two or three ion beams (masses) of interest and to repeated measurements of sample and standard gases during a single isotope ratio determination. A simplified diagram of a gas isotope ratio mass spectrometer is shown in Fig. 1. The instrument consists of six basic components: (1) a dual-inlet system for handling sample and standard gases separately; (2) an ion source where gases are ionized by a hot filament; (3) a curved flight tube located in a magnetic field that resolves ions of different masses; (4) a set of Faraday cup detectors and amplifiers for collecting and amplifying the resolved ion beams; (5) a pumping system that can maintain a

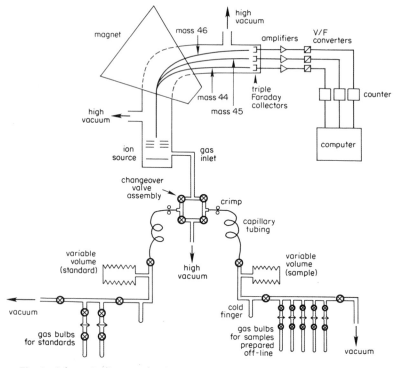

Fig. 1 Schematic diagram of an isotope ratio mass spectrometer. Circles with **X**s represent valves.

vacuum of approximately 10^{-8} torr in the flight tube during gas analysis, and 10^{-3} torr in the gas inlet system; and (6) a computer system for data acquisition and instrument control. Most commercially available isotope ratio mass spectrometers generally cover the mass range from 1 to 100 and can be used to measure not only $^{13}C/^{12}C$ ratios but also $^2H/^1H$, $^{15}N/^{14}N$, $^{18}O/^{16}O$, and $^{34}S/^{32}S$ ratios.

Only a brief summary of mass spectrometric measurements will be presented here. First, the element of interest must be isolated from the sample matrix and converted to a gas that is stable and unreactive at room temperature. In the case of carbon, samples are analyzed as CO_2. Details of sample preparation are presented later. The CO_2 gas generated from the sample is admitted to the sample side of the dual-inlet system, and a reference gas of known carbon isotope composition is admitted to the standard side of the inlet. Both gases flow through viscous capillary leaks (which prevent diffusive fractionation of the gases) up to the changeover valve assembly (Fig. 1). The configuration of these valves allows one of the gases to flow into the ion source of the mass spectrometer while the other gas is being pumped away at a comparable rate. Thus, sample and standard gases can be alternately admitted to the ion source of the mass spectrometer, and the isotopic composition of the sample can be repeatedly and rapidly compared with that of the standard. Because the ratio of ion beam currents can vary with gas pressure in the ion source, sample and standard gas pressures are matched closely by means of variable volume bellows (Fig. 1) prior to initiating isotopic measurements.

In the ion source, CO_2 molecules are ionized by an electron beam generated by a heated tungsten filament, and an accelerating potential propels the positive ions down the analyzer tube into the curved sector located in a magnetic field. Holding the strength of the magnetic field and the accelerating potential constant, the trajectory of the ions passing through the magnetic field becomes a function of the mass and energy of the ions. Because deflected to a greater extent than ions of mass 45 ($^{13}C^{16}O^{16}O, ^{12}C^{17}O^{16}O$) or mass 46 ($^{12}C^{18}O^{16}O$). Thus, the major species of CO_2 are resolved into three separate ion beams on the basis of their mass, each striking separate Faraday cup collectors (Fig. 1). The equations of motion of ions in a mass spectrometer are derived and presented elsewhere (e.g., Bowen, 1988). The neutralization of these ions upon striking the collectors results in electrical currents, which are amplified and used to compute the stable carbon isotope ratios. The $^{13}C/^{12}C$ ratio is computed from a combination of the mass 45/44 and mass 46/44 ratios and corrected for minor ^{17}O contributions to the mass 45 signal (Craig, 1957; Santrock *et al.*, 1985). Approximately 5 to 10 measurements are made on both sample and standard gases before the isotopic composition of the sample is computed. The time required to analyze a sample for $^{13}C/^{12}C$ is approximately 10 to 15 minutes for most instruments. More detailed discussions of carbon isotope measurement by isotope ratio mass spectrometry have been presented by Craig (1957), Deines (1970), and Mook and Grootes (1973). Additional information on the basic principles underlying mass spectrometry can be found in Inghram and Hayden (1954) and Roboz (1968).

III. UNITS OF MEASUREMENT AND THEIR RELATIONSHIPS

It should be emphasized that the quantity measured is not the absolute isotope ratio of the sample, but the relative difference between the isotope ratios of the sample and standard gases. As a result, a differential notation known as the delta (δ) notation (McKinney *et al.*, 1950) has been adopted for expressing relative differences in stable carbon isotope ratios between samples and standards. The $\delta^{13}C$ value is calculated from the measured carbon isotope ratios of the sample and standard gases as:

$$\delta^{13}C\ (\permil) = \left[\frac{R_{sample} - R_{standard}}{R_{standard}} \right] \times 10^3 \tag{1}$$

where $\delta^{13}C$ is the parts per thousand, or per mil ($\permil$), difference between the ^{13}C content of the sample and that of the standard, and R is the mass 45/44 ratio of the sample or standard gas.

By international convention, $\delta^{13}C$ values are always expressed relative to a calcium carbonate standard known as PDB. This standard was a limestone fossil of *Belemnitella americana* from the Cretaceous Pee Dee formation in South Carolina. As the basis of the PDB scale, it has been assigned a $\delta^{13}C$ value of 0 $\permil$. Its absolute $^{13}C/^{12}C$ ratio (R) has been reported to be 0.0112372 (Craig, 1957). The sign of the $\delta^{13}C$ value indicates whether the sample has a higher or lower $^{13}C/^{12}C$ ratio than PDB. For example, a sample with a $\delta^{13}C$ value of -10.0 $\permil$ has a $^{13}C/^{12}C$ ratio that is 10 parts per thousand (or one percent) lower than the PDB standard; a sample with a $\delta^{13}C$ value of $+1.5$ $\permil$ has a $^{13}C/^{12}C$ ratio 1.5 parts per thousand higher than the PDB standard. The PDB standard is no longer available, but other standards have been calibrated against PDB (see below), thus permitting researchers to continue reporting results relative to PDB.

In addition to the $\delta^{13}C_{PDB}$ value, several other indices of stable isotope abundance exist and are used frequently in isotopic mass balance calculations and to express data obtained in ^{13}C enrichment or tracer studies. The absolute ratio (R) of a sample is defined by rearrangement of Eq. (1) as:

$$R_{sample} = {}^{13}C/{}^{12}C = \left[\frac{\delta^{13}C}{1000} + 1 \right] \times R_{PDB} \tag{2}$$

where $R_{PDB} = 0.0112372$. The fractional abundance (F) is extremely useful in isotopic mass calculations and is related to R by the equation:

$$F = \frac{{}^{13}C}{{}^{13}C + {}^{12}C} = \frac{R}{R + 1} \tag{3}$$

Atom % is frequently used to express isotopic enrichment in samples highly enriched in ${}^{13}C$:

$$\text{atom } \% = F \times 100. \tag{4}$$

A similar index is atom % excess, which is the enrichment level of a sample following administration of a ${}^{13}C$ tracer in excess of the ${}^{13}C$ background or baseline level prior to administration of the tracer:

$$\text{atom } \% \text{ excess} = (F_{postdose} - F_{baseline}) \times 100. \tag{5}$$

All of these indices can be applied to stable isotopes other than those of carbon. Additional details on these indices and their relationship with $\delta^{13}C_{PDB}$ are provided by Hayes (1982, 1983).

IV. STABLE CARBON ISOTOPE STANDARDS

As mentioned above, the PDB standard no longer exists. However, several other primary standards were calibrated against it, so that it is still possible to express $\delta^{13}C$ values relative to PDB. Isotope ratios that have been measured against another standard can be related to PDB by the equation:

$$\delta_{(x-PDB)} = \delta_{(x-B)} + \delta_{(B-PDB)} + 10^{-3} \delta_{(x-B)}\delta_{(B-PDB)} \tag{6}$$

where all δ values have units of ‰, $\delta_{(x-PDB)}$ is the isotope ratio of the sample relative to PDB, $\delta_{(x-B)}$ is the isotope ratio of the sample relative to some standard B, and $\delta_{(B-PDB)}$ is the isotope ratio of the standard B relative to PDB (Craig, 1957).

Standards that have been calibrated to the PDB standard are distributed by the U.S. National Institute of Standards and Technology (NIST, formerly National Bureau of Standards or NBS) in Washington, D.C., and by the International Atomic Energy Agency (IAEA) in Vienna, Austria. Values for these standards are not officially certified by either organization but are arrived at by consensus (i.e., average of values obtained from many laboratories) of the geochemical community. A list of primary standards presently available and their $\delta^{13}C_{PDB}$ values are shown in Table 1. Note that some variation exists in the values reported for these standards. Some of these standards can also be used for calibration of $\delta^{18}O$ results.

Primary carbon isotope standards (Table 1) should not be used for routine sample analysis because their availability is limited, and they can be fraction-

Table 1

Stable Carbon Isotope Standards Available from
NIST and IAEA and Their $\delta^{13}C_{PDB}$ Values

Standard	Description	$\delta^{13}C$(‰ vs PDB)
NBS-16	CO_2	-41.48^a
		-41.64^b
		-41.59^c
NBS-17	CO_2	-4.41^a
		-4.48^b
		-4.45^c
NBS-18	$CaCO_3$	-5.00^a
		-5.04^b
		-5.04^c
NBS-19	$CaCO_3$	$+1.92^a$
		$+1.96^b$
NBS-20	$CaCO_3$	-1.06^a
		-1.08^c
NBS-21	Graphite	-28.10^a
		-28.16^c
NBS-22	Oil	-29.63^a
		-29.73^c
		-29.81^d
ANU sucrose	Sucrose	-10.40^b
		-10.47^c
PEF-1	Polyethylene foil	-31.60^b
		-31.77^c

[a] Coplen *et al.* (1983).
[b] Gonfiantini (1984).
[c] Hut (1987).
[d] Schoell *et al.* (1983).

ated and/or contaminated by regular use. Instead, these primary standards should be used to calibrate secondary working standards, which can be used routinely.

There are two basic approaches to routine use of working standards. The first approach involves preparation of approximately 20-ml of CO_2 from a working standard such as glucose, $CaCO_3$, or corn oil. After carefully checking the isotopic composition against other working standards, the gas can then be used for routine work. At the end of a day's run, the working standard can be retrieved from the mass spectrometer inlet system and frozen back into the gas bottle using liquid nitrogen, and used repeatedly until the volume becomes too small for continued use. A 20-ml working standard should be large enough to run 60 samples/day for approximately 3–4 weeks. The isotopic composition of this working standard should be checked by running it against another working standard prior to each day's work. The disadvantage of this approach is that a new working standard must be prepared approximately once each month.

The second approach involves the use of a lecture bottle or cylinder of compressed, high-purity (99.999%) CO_2 as a working standard. This approach is very popular because it involves no preparation of CO_2, and because a cylinder of CO_2 is a virtually inexhaustible supply of standard gas. While it is possible to connect a CO_2 cylinder to the standard side of the mass spectrometer inlet system and admit small quantities of gas directly, many investigators prefer to fill smaller gas bottles off-line at greatly reduced pressure. Accidental pressurization of the gas inlet system and/or analyzer tube of a mass spectrometer can cause serious structural damage, so extreme caution should be exercised if a pressurized cylinder is attached directly to the mass spectrometer. The isotopic composition of the tank CO_2 should be checked daily against another working standard to ensure that no drift has occurred. The aliquot of working standard used for each day's run is usually pumped away at the end of the day because of the large supply of CO_2 working standard available. To avoid potential isotopic fractionation between liquid and gaseous phase CO_2, pressure of CO_2 in the cylinder should be less than 600 psi to eliminate the possibility of a liquid phase at room temperature. While most compressed CO_2 tends to have low $\delta^{13}C$ values (usually -30 to -45 ‰ vs PDB), it is possible to special order CO_2 with a specified $\delta^{13}C$ value.

V. SAMPLE PREPARATION TECHNIQUES

Because of instrumental requirements, carbon must be converted to CO_2 for stable isotope ratio measurements. Any method for converting organic or inorganic forms of carbon to CO_2 must not cause isotopic fractionation of the sample, should produce quantitative yields of carbon to permit determination of percent carbon in the sample, and should be rapid and inexpensive.

Most of the error associated with isotopic measurements results from sample preparation. Mass spectrometer precision (1 SD), as determined by repeated analyses of the same gas sample, is often as low as 0.01 ‰. By contrast, different preparations of aliquots of the same sample will generally have a precision (1 SD) of 0.1 ‰.

The total carbon present in a complex sample (e.g., leaf tissue) can be analyzed for $\delta^{13}C$. In many studies, however, only a specific biochemical fraction (e.g., lipids), or a specific intramolecular position may be of interest. Thus, the first step is often the isolation of the chemical fraction of interest from the complex sample matrix. Care should be taken at this stage to ensure that the methods of isolation do not result in isotopic alteration of the sample. Isotopic fractionation of a chemical end product resulting from extraction and isolation procedures can be detected by subjecting a quantity of the desired pure end product (obtained commercially) to isotopic analysis. Following isotopic characterization of that material, it should then be submitted to the intended isolation and purification procedure, and its isotopic composition redetermined. A difference in the isotopic composition before and after the analytical procedure would indicate that the procedure alters the isotope ratio of the chemical of interest. When present, this isotopic alteration will typically result from failure to recover all of the starting material (incomplete yield), or the presence of foreign organic matter such as solvents or other reagents used during the procedure. If the isotopic alteration is consistent and repeatable, isotopic ratios of materials subjected to the procedure in question can be corrected. However, if the fractionation is variable, additional effort to increase yield and purity of the final product will be necessary. Specific examples of the general approach outlined above can be found in Abelson and Hoering (1961) and Blair *et al.* (1985).

The amount of sample required is a direct function of the carbon content of the sample, the volume of the mass spectrometer inlet system, the ionization efficiency of the ion source, and the degree of precision required. Thus, sample size will vary from laboratory to laboratory. Hayes *et al.* (1978) have calculated a theoretical minimum sample size of 0.3 nmoles of carbon, assuming that the efficiency of the ion source was 10^{-4} ions/molecule and that a precision of 0.1 ‰ was needed. While efforts are being made to reduce the minimum sample size required, present sample size requirements are several orders of magnitude higher. For example, most mass spectrometers now being manufactured have minimum requirements of approximately 10 μl of CO_2 at STP, corresponding to about 5 μg of carbon. In practice, most laboratories use approximately 0.5 to 5 mg of carbon for conversion to CO_2. When samples are isotopically heterogeneous (soils, sediments, gut contents, etc.), it is advisable to grind and mix the sample thoroughly and use larger samples to avoid sampling error, which can result in reduced precision and accuracy of isotopic measurements.

A variety of sample types are encountered in biological stable isotope applications, each requiring a different technique for sample preparation. These sample types can be broadly classified into three categories: (1) organically bound carbon; (2) carbonate carbon; and (3) CO_2 gas (e.g., respiratory CO_2, atmospheric CO_2, soil CO_2, dissolved CO_2). In this chapter, only organic and carbonate carbon techniques will be addressed, while techniques for purifying CO_2 from air and respiratory gases will be covered in Chapters 12 and 14 of this volume, respectively.

A. Organic Samples

Conversion of organic samples to CO_2 for isotopic analysis is accomplished by dry combustion in an excess of oxygen. Although a variety of combustion techniques have been described, the simplest and fastest involve the combustion of individual samples in sealed, evacuated quartz or vycor tubes containing CuO as the oxygen source (Buchanan and Corcoran, 1957; Frazer and Crawford, 1963). Because each sample is prepared in its own container, there is no chance for memory effects. This technique will be emphasized here.

Organic samples should be dried thoroughly prior to combustion to reduce the amount of water vapor generated during combustion and to achieve greater accuracy in percent carbon determinations (by obtaining more accurate sample weight) if they are to be performed. Failure to dry samples adequately may result in explosion of tubes during combustion due to the generation of large amounts of water vapor. Large, complex samples (e.g., leaf litter, elephant feces) should be ground to a powder and thoroughly mixed to reduce the probability of sampling error when a few milligrams are subsampled for combustion. Samples suspected of containing carbonate carbon should be treated with dilute HCl (1N) and rinsed with distilled water prior to drying. The cessation of bubbling indicates that carbonates have been removed, and can be verified under a dissecting microscope while the sample is in dilute HCl. Carbonate carbon is often more enriched in ^{13}C than organic carbon and may confound analyses of organic matter if it is not removed. It should be noted that treatment of organic matter with dilute HCl may result in some loss of organic carbon (Showers and Angle, 1986), and that failure to rinse HCl from the sample may result in cracking of the quartz tube during combustion. Dilute H_3PO_4 has been suggested as an alternative for removing carbonates from organic samples (Showers and Angle, 1986).

Combustion tubes are prepared by cutting quartz or vycor tubing (9 mm o.d. $\times$ 7 mm i.d.) into 15–20 cm lengths and sealing each tube at one end with a gas/oxygen flame. Goggles with shade 8 or 9 welding lenses should be worn whenever heating quartz or vycor to the softening point. Standard glassblowing glasses or goggles are inadequate for quartz work. Furthermore, all quartz or vycor work should be done under a fume hood or in a well-ventilated room because small quantities of ozone and other toxic gases are emitted when they are heated. Sample boats approximately 3 cm in length are cut from 6-mm o.d. $\times$ 4-mm i.d. quartz or vycor tubing and sealed at one end. Combustion tubes and sample boats are then heated in a muffle furnace for 1 hr at 900°C in order to remove potential organic contaminants. Sample boats can be stored in a clean glass jar until ready for use.

The reagents required for the combustions are wire form CuO and reduced copper. Wire form CuO is placed in a crucible and heated in a muffle furnace at 800°C for 3 hr to remove organic contamination. Commercially available CuO suitable for microanalysis contains approximately 0.3 μmol C/g CuO; after heating, the carbon content is approximately 0.1–0.3 μmol C/g CuO (Wedeking *et al.*, 1983). Upon cooling, batches of 10–100 combustion tubes are loaded with 1 g of purified CuO using a long-stem funnel, and stored in a desiccator until needed.

As mentioned earlier, the amount of sample required will depend largely on the volume of the mass spectrometer inlet system and the carbon content of the sample. For most mass spectrometers currently in use, sample sizes of 5 mg for plant tissue ($\sim$42% carbon), 3 mg for animal tissue ($\sim$50–70% carbon, depending on lipid content), and 500 mg for soils and sediments ($\sim$1–3% carbon) should prove adequate. Too much sample (e.g., 20–30 mg of plant tissue) may result in explosion of the sample tube. If the percent carbon is to be determined, an analytical balance readable to at least 0.1 mg should be used for weighing the sample into the boat. The sample boat should be handled only with clean forceps to avoid contamination from body oils. After the sample has been weighed into the boat, the boat should be placed open end first into a combustion tube held horizontally. Using a glass rod, the boat can then be pushed slowly into the combustion tube until it contacts the CuO. At this point, the sample can be mixed with the CuO by gentle shaking. Using a long stem funnel, approximately 0.5 g of reduced granular copper ($\sim$20 mesh) should be added to the combustion tube on top of the sample/CuO mixture. Tubes should be labeled with an engraving tool or a high-temperature marking pencil.

Combustion tubes loaded with samples are attached to a tube-sealing manifold (Fig. 2a) with Cajon ultratorr o-ring fittings. When first exposing the sample tubes to the vacuum, valve 2 should be opened very slowly to avoid sucking the samples out of the tubes. The liquid nitrogen trap (Fig. 2a) helps pump water vapor out of the manifold and freezes any oil vapor that migrates from the vacuum pump. When a vacuum of $<10^{-2}$ torr has been achieved, the combustion tubes can be sealed with a gas/oxygen torch. Sealed tubes are again shaken in order to promote contact between the sample and the reagents, thus ensuring complete combustion.

Inside the muffle furnace, each sealed tube is placed inside a ceramic or inconel tube (1.3 cm i.d.) to shield it from the effects of an exploding tube. The muffle furnace is then heated to 900°C and the temperature maintained for 2 hr. At temperatures above 500°C, the CuO undergoes pyrolytic decomposition, providing O_2 for oxidation of organic carbon to CO_2. After 2 hr, the muffle furnace is slowly cooled (1°/min) to 650°C, where the temperature is maintained for an additional 2 hr before cooling to room

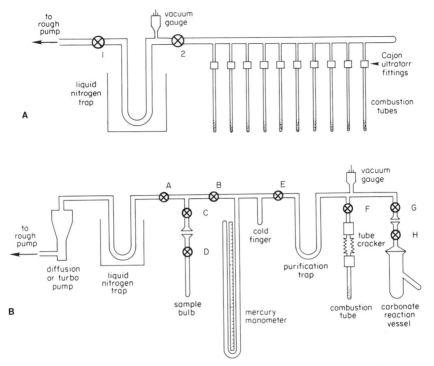

Fig. 2 Vacuum systems for sealing quartz combustion tubes under vacuum (A), and for purification and measurement of CO_2 generated by combustion of organic matter or acidification of carbonate (B). Circles with Xs represent valves.

temperature. While muffle furnace temperatures can be controlled manually, a programmable muffle furnace is ideal for this application. During the 650°C dwell period, the reduced copper eliminates halogens, and catalyzes the conversion of any CO to CO_2, NO_x to N_2, and SO_x to $CuSO_4$ (Frazer and Crawford, 1963). The prolonged cool-down period also promotes reoxidation of the copper, thereby eliminating any excess oxygen (Sakai *et al.*, 1976). High resolution mass spectrometric analysis of the gases produced by combustion of a variety of organic compounds by this method revealed the presence of only CO_2, H_2O, and N_2 (Boutton *et al.*, 1983). Sealed tubes that have been combusted should not be stored for more than 5 days prior to isolation and purification of the CO_2 because carbonate forms slowly during storage, and CO_2 becomes more depleted by 1 to 3 ‰ after two weeks (Stuiver *et al.*, 1984; Engel and Maynard, 1989). However, in the event that samples cannot be processed within 5 days, this isotopic depletion can be overcome by recombusting the tubes prior to the sample purification procedure outlined below (Engel and Maynard, 1989).

Next, the CO_2 must be separated from the other combustion products by cryogenic distillation. A simple vacuum line designed for this purpose is illustrated in Fig. 2b and consists of a mechanical vacuum pump, a turbo-molecular or diffusion pump to provide a high vacuum ($\sim 10^{-6}$ torr), a tube cracker for breaking the combustion tube under vacuum, a purification trap for cryogenic distillation of the combustion products, a manometer for measuring the volume of CO_2 produced, and a port for attaching sample bulbs.

During the CO_2 isolation and purification procedure, safety glasses should be worn to prevent injury in the event that a combustion tube explodes or a Dewar of liquid nitrogen or ethanol-dry ice slush implodes. The combustion tube is scored at one end and inserted into the tube cracker (DesMarais and Hayes, 1976), a sample bulb is attached to the manifold, all valves are opened, and the entire system is pumped down to approximately 10^{-4} torr or lower. Then, a liquid nitrogen Dewar ($-196°C$) is placed on the purification trap, and valve E is closed (Fig. 2b). The combustion tube is broken, releasing the gases into the vacuum line; water and CO_2 are allowed to freeze into the purification trap. After 4 min, valve F is closed, valve E is opened, and all noncondensible gases are pumped away. When the vacuum gauge reaches 10^{-3} torr or lower, valve B is closed and the liquid nitrogen Dewar is removed from the purification trap and replaced with a Dewar containing an ethanol-dry ice slush ($-86°C$). This allows the CO_2 to sublime but keeps the water frozen in the trap. To measure the volume of CO_2 produced, the CO_2 is transferred into the manometer cold finger by cooling it with liquid nitrogen. The progress of this transfer can be monitored with the thermocouple vacuum gauge and should take approximately 2 min, depending on the volume of CO_2 produced and the internal diameter of the vacuum system. When the transfer is complete, valves B and E are closed, the liquid nitrogen Dewar is removed from the cold finger, and the CO_2 is allowed to expand into the mercury manometer previously calibrated with known volumes of CO_2. After noting the manometer reading, valve A is closed and valves B, C, and D are opened. The CO_2 is finally transferred into the sample bulb by immersing it in liquid nitrogen for approximately 2 min. When the transfer is complete, valves D and C are closed, and the sample bulb can be detached from the purification system and attached to the inlet of the mass spectrometer (Fig. 1).

A similar technique has been described for combusting organic samples in pyrex tubes at 550°C (Sofer, 1980). This technique is an attractive alternative due to the lower cost of pyrex relative to quartz tubing. However, combustions at 550°C in pyrex tubes have been shown to result in incomplete combustion and more variable $\delta^{13}C$ values relative to samples combusted in quartz tubes at 800 to 900°C (Boutton *et al.*, 1983; LeFeuvre and Jones, 1988). Incomplete combustion will eliminate the possibility of determining percent carbon on a sample. Intimate contact between sample and CuO are critical for combustions in pyrex tubes at 550°C, and caution is urged when combusting organic samples in pyrex tubes at 550°C.

The importance of manometric measurements during sample preparation should be emphasized. In addition to permitting calculation of percent carbon in the sample, these measurements can provide quality control information. For example, if samples with a known percent carbon are being prepared, failure to obtain the known value suggests that an error occurred at some point during sample preparation, and extreme caution should be exercised in interpreting the isotope ratio obtained from that sample. A value for percent carbon higher than expected might indicate the presence of a contaminant such as water vapor; a value lower than expected might indicate incomplete combustion of the sample or incomplete recovery of CO_2 during cryogenic purification.

It should be noted that an electronic capacitance manometer could be substituted for the mercury manometer shown in Fig. 1. Although it is more expensive, the capacitance manometer eliminates the need for mercury, which is highly toxic, and also affords significantly greater accuracy for determination of percent carbon. Accuracy for electronic manometers ranges from 0.15 to 1% of reading, depending on the model, while that for a mercury manometer ranges from approximately 2 to 5% of reading.

B. Carbonate Carbon

Carbonate carbon is converted to CO_2 by reaction under vacuum with concentrated phosphoric acid. Because the oxygen in the CO_2 generated by carbonate acidification is derived from the carbonate, the CO_2 derived from this procedure can be used to measure both the $\delta^{13}C$ and $\delta^{18}O$ of the carbonate sample, provided the reaction temperature remains known and constant and water-free 100% phosphoric acid is used for acidification. If only the $\delta^{13}C$ value of the carbonate is required or if the carbonate is in aqueous solution, 85% phosphoric acid is suitable. However, if the $\delta^{18}O$ of the carbonate is to be measured simultaneously with $\delta^{13}C$ value, the reaction must be carried out with 100% phosphoric acid to avoid oxygen exchange between water and CO_2.

Several methods have been described for preparing 100% phosphoric acid, although the method outlined by Bowen (1966) is the most commonly used. Briefly, 1.4 kg of P_2O_5 is added slowly to 3.2 kg of 85% H_3PO_4 while stirring. Approximately 1 g of CrO_3 is added to oxidize any organic contaminants, and the solution is heated at 200°C for 7 hr to drive off water. Then, 3 ml of H_2O_2 is added to reduce any remaining CrO_3, and heating is continued for another 4.5 hr at 220°C before cooling to room temperature. The 100% phosphoric acid should be green and have a density of at least 1.9. This acid is extremely hygroscopic, and must be stored carefully to avoid contamination by atmospheric water vapor. A glass-stoppered reagent bottle is suitable for this purpose.

If organic matter is present in the carbonate sample, it should be removed by roasting under vacuum at approximately 380°C for 1 hr. Alternatively, organic matter can be removed from the carbonate by soaking the sample overnight in 5% sodium hypochlorite solution (Grossman and Ku, 1986). Failure to remove organic matter can influence $\delta^{13}C$ measurements of carbonates.

For pure, dried $CaCO_3$, a 10-mg sample will generate 2.45 ml of CO_2 at 25°C, which should be adequate for isotopic analysis on most mass spectrometers. However, in many cases (e.g., soils), the sample will not be pure $CaCO_3$, and an adequate sample size will need to be determined by trial and error.

Acidification of the carbonate sample is carried out in a reaction vessel in which the phosphoric acid can be kept separate from the sample until the vessel is evacuated. A typical reaction vessel is shown in Fig. 2b and consists of two separate parts. The top portion includes a vacuum stopcock and can be attached to the sample purification system (Fig. 2b), while the bottom portion includes a main chamber for the carbonate sample and a side arm for the 100% phosphoric acid. The upper and lower portions can be joined together by means of a ground-glass or o-ring joint.

The sample is weighed into the main portion of the reaction vessel, keeping sample off the side walls and out of the side arm. An excess of 100% phosphoric acid (~ 3 ml) is placed in the side arm. Any acid present in the sample compartment could contact the sample prematurely and ruin the sample. When sample and acid are both loaded, the upper and lower portions of the reaction vessel can be joined and the complete assembly attached to the vacuum system, and evacuated to 10^{-3} torr or lower (Fig. 2b). During this time, the phosphoric acid may bubble as dissolved gas escapes. If the carbonate sample is an aqueous solution, it should be frozen with an ethanol-dry ice slush prior to evacuation of the reaction vessel. When the vessel has been pumped to 10^{-3} torr or lower, valves H and G are closed (Fig. 2b) and the vessel is removed from the system and placed in a water bath at 25°C for 30 min.

After 30 min, the reaction vessel is removed from the water bath and the phosphoric acid is tipped into the sample chamber. The walls of the sample chamber should be thoroughly coated with acid in the event that some sample adhered to the walls during loading. The reaction vessel is then returned to the water bath and allowed to react at 25°C overnight. The following day, the reaction vessel is removed from the water bath, and again the walls of the sample chamber should be coated with acid to ensure that all of the sample has been acidified. The reaction vessel can then be attached to the vacuum system in Fig. 2b and the CO_2 purified, measured manometrically (if necessary), and transferred into a sample bulb following the same procedure outlined above for purification of combusted organic samples. The carbonate samples, especially those reacted with 100% phosphoric acid, should contain far less water vapor and noncondensible gases than the combusted samples. Solid carbonates can also be decomposed to CO_2 for $\delta^{13}C$ analysis (but not $\delta^{18}O$) by mixing the sample with $Na(PO_3)_n$ in a vycore tube, and then heating the mixture on a vacuum line (B. Fry, personal communication).

VI. PROSPECTS FOR THE FUTURE

Perhaps the major area of interest in isotope ratio mass spectrometry is the development of on-line sample preparation systems that can be automated and interfaced with the mass spectrometer, allowing unattended sample preparation and determination of $\delta^{13}C$. In the case of organic carbon samples, three options are commercially available. In the first option, samples are combusted in an elemental analyzer, the CO_2 isolated from other combustion products by gas chromatography, and the helium carrier gas sweeps the CO_2 into the ion source of the double- or triple-collecting isotope ratio mass spectrometer (Preston and Owens, 1985; Barrie and Lemley, 1989). In this configuration, there is no dual inlet for rapid comparison between sample and standard gases, so precision (1 SD = 0.6 ‰) is poorer than that attainable by conventional isotope ratio mass spectrometry. This precision is marginal or unacceptable for most natural abundance work but is suitable for high enrichment tracer work.

In the second option, samples are again combusted in an elemental analyzer and the CO_2 separated from the other combustion products by gas chromatography, but the CO_2 is then removed from the carrier gas by passing the gas stream through a liquid nitrogen trap. When the sample has been collected completely, the pure CO_2 is thawed and admitted into the dual inlet of a conventional isotope ratio mass spectrometer. Because sample and standard are compared in the conventional manner, precision attainable with this configuration is approximately 0.1 ‰ or better, similar to that achieved with sealed-tube combustions. In the third option, a combustion furnace has been placed in line between a gas chromatograph and an isotope ratio mass spectrometer, allowing $\delta^{13}C$ measurements to be made on CO_2 from specific organic compounds (Barrie *et al.*, 1984; Freedman *et al.*, 1988). Again, there is no dual-inlet system and the CO_2 is swept into the ion source by a carrier gas, so precision is approximately 1.0 ‰. However, this precision is achieved with only 0.18 μl of CO_2, allowing isotope ratios to be measured on extremely small quantities of specific organic compounds.

ACKNOWLEDGMENTS

This manuscript benefited from reviews by David D. Briske, Ethan L. Grossman, Peter D. Klein, and Brian Fry. The assistance of Sylvia Dudash with manuscript preparation is gratefully acknowledged. Support for the preparation of this manuscript was provided by project H-6945 of the Texas Agricultural Experiment Station.

REFERENCES

Abelson, P. H., and Hoering, T. C. (1961). Carbon isotope fractionation in formation of amino acids by photosynthetic organisms. *Proc. Nat. Acad. Sci.* **47**, 623–632.

Barrie, A., and Lemley, M. (1989). Automated $^{15}N/^{13}C$ analysis of biological materials. *Am. Lab.* **21**(8), 54–63.

Barrie, A., Bricout, J., and Koziet, J. (1984). Gas chromatography-stable isotope ratio analysis at natural abundance levels. *Biomed. Mass Spectrom.* **11**, 583–588.

Blair, N., Leu, A., Munoz, E., Olsen, J., Kwong, E., and DesMarais, D. (1985). Carbon isotopic fractionation in heterotrophic microbial metabolism. *Appl. Environ. Microbiol.* **50**, 996–1001.

Boutton, T. W., Wong, W. W., Hachey, D. L., Lee, L. S., Cabrera, M. P., and Klein, P. D. (1983). Comparison of quartz and pyrex tubes for combustion of organic samples for stable carbon isotope analysis. *Anal. Chem.* **55**, 1832–1833.

Bowen, R. (1966) "Paleotemperature Analysis." Elsevier Publishing Company, New York.

Bowen, R. (1988). "Isotopes in the Earth Sciences." Elsevier Applied Science, New York.

Buchanan, D. L., and Corcoran, B. J. (1957). Sealed tube combustions for the determination of carbon-14 and total carbon. *Anal. Chem.* **31**, 1635–1638.

Coplen, T. B., Kendall, C., and Hopple, J. (1983). Comparison of stable isotope reference samples. *Nature (London)* **302**, 236–238.

Craig, H. (1957). Isotopic standards for carbon and oxygen and correction factors for mass spectrometric analysis of carbon dioxide. *Geochim. Cosmochim. Acta* **12**, 133–149.

Deines, P. (1970). Mass spectrometer correction factors for the determination of small isotopic composition variations of carbon and oxygen. *Int. J. Mass Spectrom. Ion Phys.* **4**, 283–295.

DesMarais, D., and Hayes, J. M. (1976). Tube cracker for opening glass-sealed ampoules under vacuum. *Anal. Chem.* **48**, 1651–1652.

Engel, M. H., and Maynard, R. J. (1989). Preparation of organic matter for stable carbon isotope analysis by sealed tube combustion: A cautionary note. *Anal. Chem.* **61**, 1996–1998.

Frazer, J. W., and Crawford, R. (1963). Modifications in the simultaneous determination of carbon, hydrogen, and nitrogen. *Mikrochim. Acta* **1963/3**, 561–566.

Freedman, P. A., Gillyon, E. C. P., and Jumeau, E. J. (1988). Design and application of a new instrument for GC-isotope ratio MS. *Am. Lab* **20**(6), 114–119.

Gonfiantini, R. (1984). "Advisory Group Meeting on Stable Isotope Reference Samples for Geochemical and Hydrological Investigations." Report to the Director General, IAEA, Vienna.

Grossman, E. L., and Ku, T. (1986). Oxygen and carbon isotope fractionation in biogenic aragonite: Temperature effects. *Chem. Geol. (Isotope Geosciences Section)* 59, 59–74.

Hayes, J. M. (1982). Fractionation, *et al.*: An introduction to isotopic measurements and terminology. *Spectra* 8, 3–8.

Hayes, J. M. (1983). Practice and principles of isotopic measurements in organic geochemistry. In "Organic Geochemistry of Contemporaneous and Ancient Sediments" (W. G. Meinschein, ed.), pp. 5-1–5-31. Society for Economic Paleontologists and Mineralogists, Bloomington, Indiana.

Hayes, J. M., DesMarais, D. J., Peterson, D. W., Schoeller, D. A., and Taylor, S. P. (1978). High precision stable isotope ratios from microgram samples. *Adv. Mass Spectrom.* 7A, 475–480.

Hut, G. (1987). "Consultants' Group Meeting on Stable Isotope Reference Samples for Geochemical and Hydrological Investigations." Report to the Director General, IAEA, Vienna.

Inghram, M. G., and Hayden, R. J. (1954). "Mass Spectroscopy." Nuclear Science Series, Report Number 14. National Academy of Sciences—National Research Council, Washington, D.C.

LeFeuvre, R. P., and Jones, R. J. (1988). Static combustion of biological samples sealed in glass tubes as a preparation for $\delta^{13}C$ determination. *Analyst* 113, 817–823.

McKinney, C. R., McCrea, J. M., Epstein, S., Allen, H. A., and Urey, H. C. (1950). Improvements in mass spectrometers for the measurement of small differences in isotope abundance ratios. *Rev. Sci. Instrum.* 21, 724–730.

Mook, W. G., and Grootes, P. M. (1973). The measuring procedure and corrections for the high-precision mass-spectrometric analysis of isotopic abundance ratios, especially referring to carbon, oxygen, and nitrogen. *Int. J. Mass Spectrom. Ion Physics* 12, 273–298.

Preston, T., and Owens, N. J. P. (1985). Preliminary ^{13}C measurements using a gas chromatograph interfaced to an isotope ratio mass spectrometer. *Biomed. Mass Spectrom.* 12, 510–513.

Roboz, J. (1968). "Introduction to Mass Spectrometry. Instrumentation and Techniques." Interscience Publ., New York.

Sakai, H., Smith, J. W., Kaplan, I. R., and Petrowski, C. (1976). Micro-determinations of C, N, S, H, He, metallic Fe, $\delta^{13}C$, $\delta^{15}N$, and $\delta^{34}S$ in geologic samples. *Geochem. J.* 10, 85–96.

Santrock, J., Studley, S. A., and Hayes, J. M. (1985). Isotopic analyses based on the mass spectrum of carbon dioxide. *Anal. Chem.* 57, 1444–1448.

Schoell, M., Faber, E., and Coleman, M. (1983). Carbon and hydrogen isotopic compositions of the NBS-22 and NBS-21 stable isotope reference materials: An interlaboratory comparison. *Org. Geochem.* 5, 3–6.

Showers, W. J., and Angle, D. G. (1986). Stable isotopic characterization of organic carbon accumulation on the Amazon continental shelf. *Cont. Shelf Res.* 6, 227–244.

Sofer, Z. (1980). Preparation of carbon dioxide for stable carbon isotope analysis of petroleum fractions. *Anal. Chem.* 52, 1389–1391.

Stuiver, M., Burk, R., and Quay, P. (1984). $^{13}C/^{12}C$ ratios in tree rings and the transfer of biospheric carbon to the atmosphere. *J. Geophys. Res.* 89, 11731–11748.

Wedeking, K. W., Hayes, J. W., and Matzigkeit, U. (1983). Procedures of organic geochemical analysis. In "Earth's Earliest Biosphere: Its Origin and Evolution" (J. W. Schopf, ed.), pp. 428–441. Princeton University Press, Princeton, New Jersey.

11

Stable Carbon Isotope Ratios of Natural Materials: II. Atmospheric, Terrestrial, Marine, and Freshwater Environments

Thomas W. Boutton

Department of Rangeland Ecology and Management
Texas Agricultural Experiment Station
Texas A&M University
College Station, Texas 77843

I. INTRODUCTION

Stable carbon isotope ratios of most natural materials of biological interest range from approximately 0 to -110 ‰ versus PDB. This variation is a result of isotopic fractionation during physical, chemical, and biological processes. Inorganic carbon in seawater, freshwater, and carbonates is relatively enriched in ^{13}C, and the isotopic composition of these materials is determined largely by isotope exchange between the atmosphere and hydrosphere. By contrast, organic carbon is generally depleted in ^{13}C as a result of biological isotope fractionation, occurring primarily during the photosynthetic process. This chapter will summarize the carbon isotopic composition of major components of the atmospheric, terrestrial, marine, and freshwater environments. Microorganisms are excluded because few measurements have been made on natural populations, and animals will be addressed in Chapter 13, this volume. The data presented are highly generalized and abstracted mostly from several detailed reviews on the carbon isotope composition of natural materials (Degens, 1969; Deines, 1980; Anderson and Arthur, 1983; Schidlowski *et al.,* 1983; Galimov, 1985; Bowen, 1988; O'Leary, 1981, 1988). All $\delta^{13}C$ values reported below are relative to the PDB standard.

CARBON ISOTOPE TECHNIQUES
Copyright © 1991 by Academic Press, Inc. All rights of reproduction in any form reserved.

II. ATMOSPHERIC ENVIRONMENT

Atmospheric CO_2 serves as the major link between the inorganic and organic portions of the global carbon cycle, and between terrestrial and marine ecosystems (Golubic *et al.*, 1979). It participates in equilibrium exchange reactions with the ocean carbonate system and serves as a substrate for photosynthesis. Because the ocean is a much larger carbon pool than the atmosphere, the $\delta^{13}C$ value of atmospheric CO_2 is determined largely by atmosphere–ocean exchange of CO_2. As a result of an equilibrium isotope fractionation, the $\delta^{13}C$ value of CO_2 released from the ocean into the atmosphere is approximately 9 ‰ lower than the total inorganic carbon ($\delta^{13}C \approx +1.5$ ‰) dissolved in the surface waters of the ocean (Mook, 1986). In addition, global patterns of photosynthesis and respiration, as well as human activities, have both a seasonal and a long-term effect on $\delta^{13}C$ values of atmospheric CO_2.

The mean $\delta^{13}C$ value of atmospheric CO_2 from remote areas (e.g., Levin *et al.*, 1987) is currently near -7.8 ‰ (Fig. 1) but varies seasonally in response to patterns of photosynthesis and respiration in the northern hemisphere (Mook *et al.*, 1983). In autumn, atmospheric $\delta^{13}C$ is maximal due to photosynthetic discrimination against ^{13}C by plants (see below) during the recent growing season. By contrast, $\delta^{13}C$ of atmospheric CO_2 is lowest in spring due to rapid soil respiration, which releases CO_2 derived from organic matter more depleted in ^{13}C than the atmosphere. For example, seasonal values for atmospheric CO_2 at Barrow, Alaska range from approximately -8.4 ‰ during the winter to -7.4 ‰ during late summer (Mook, 1986). Seasonal ^{13}C variation in the southern hemisphere is less pronounced and approximately 6 months out of phase with the northern hemisphere (Mook *et al.*, 1983).

In addition to seasonal variation, there has been a long-term decline in atmospheric $\delta^{13}C$ corresponding to the increase in atmospheric CO_2 concentration from approximately 280 ppm before the industrial revolution to approximately 345 ppm at present (Friedli *et al.*, 1986). Anthropogenic CO_2 is derived from combustion of fossil fuels and from deforestation. All of the fossil fuels have low $\delta^{13}C$ values, and their combustion produces CO_2 that is approximately -26 ‰ (Fig. 1), substantially lower in ^{13}C than the atmosphere (Tans, 1981). Vegetation removed during deforestation is also low in ^{13}C, so that when it decomposes or burns, CO_2 with a $\delta^{13}C$ value of approximately -27 ‰ is released into the atmosphere. Thus, the net result of fossil fuel combustion and deforestation is dilution of the ^{13}C content of the atmosphere. The $\delta^{13}C$ value of atmospheric CO_2 has decreased by approximately 1.2 ‰ over the past 130 years (Friedli *et al.*, 1986) and is

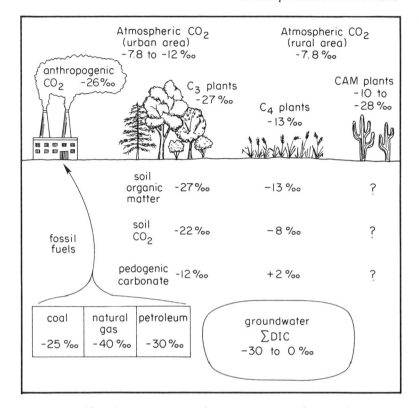

Fig. 1 Stable carbon isotope ratios of major components of terrestrial ecosystems.

continuing to decline. Isotopic exchange between atmospheric CO_2 and oceanic HCO_3^- pools dampens this anthropogenic signal.

Atmospheric CH_4 and CO represent much smaller carbon pools than CO_2, but they have important links with the biosphere and both are increasing in concentration. The atmospheric concentration of CH_4 has more than doubled in the past 200 yr from 0.7 to 1.7 ppm, and is increasing presently at 1 to 2 percent per year. The $\delta^{13}C$ of atmospheric CH_4 from remote areas 200 yr ago was approximately -50 ‰ (Craig *et al.*, 1988), while the present value is -47.7 ‰ (Lowe *et al.*, 1988). It has been suggested that this temporal shift in $\delta^{13}C$ resulted from an increase in biomass burning (deforestation), which produces CH_4 with a $\delta^{13}C$ of -25 ‰ (Craig *et al.*, 1988).

The most important sources of atmospheric CH_4 include bacterial fermentation of organic matter in anaerobic environments (wetlands, rice paddies, landfills, sediments, and the digestive systems of ruminants and termites), biomass burning, fossil fuel burning, coal fields, and natural gas. $\delta^{13}C$ values for these sources are mostly more depleted than the atmospheric pool, ranging from approximately -90 to -15 ‰ with a mean value of -56 ‰ (Craig *et al.*, 1988). The most important sink for atmospheric CH_4 is oxidation to CO by the OH radical, resulting in an approximate $+10$ ‰ shift in the isotopic composition of the remaining CH_4, thereby reconciling the isotopic difference between the relatively depleted sources and the atmospheric CH_4 pool. The relative importance and isotopic composition of each CH_4 flux into the atmosphere is difficult to measure and known poorly, but recent isotopic mass balance calculations suggest that 23 to 32% of atmospheric CH_4 is derived from fossil sources (Lowe *et al.*, 1988).

Sources, sinks, and isotopic composition of atmospheric CO are not well known. Based on ^{13}C and ^{18}O content, several distinct species of CO have been identified (Stevens *et al.*, 1972), suggesting that CO from different sources may be isotopically unique. However, the only source that has been isotopically characterized to date is automobile exhaust, with a $\delta^{13}C$ of -27.4 ‰ (Stevens *et al.*, 1972). Atmospheric CO is highly variable in concentration in space and time, with concentrations between 0.15 and 0.20 ppm in the northern hemisphere and .05 to .06 ppm in the southern hemisphere (Crutzen, 1983). The global concentration is increasing at approximately 0.8 to 1.4 percent per year.

This variability in concentration is matched by isotopic variability, with $\delta^{13}C$ values for CO from sites throughout the world ranging from approximately -32 to -22 ‰ (Stevens *et al.*, 1972). While inputs of CO from incomplete combustion of fossil fuels and from oxidation of CH_4 by OH radicals are clearly important, models of the global CO cycle suggest that biological inputs such as oxidation of biogenic hydrocarbons emitted by vegetation must be equally important (Crutzen, 1983). Additional information on the isotopic composition of CO sources could enhance our understanding of this important trace gas.

III. TERRESTRIAL ENVIRONMENT

Most of the natural isotopic variation of interest to biologists results from carbon isotope fractionation during photosynthesis (Farquhar *et al.*, 1989). Terrestrial plants can be divided into three major photosynthetic types, each with unique carbon isotope fractionation patterns (Fig. 1). The $\delta^{13}C$ of these different photosynthetic types are largely a result of: (1) the biochemical properties of the primary CO_2-fixing enzymes; and (2) limitations to CO_2 diffusion into the leaf (O'Leary, 1988). Plants with the C_3 pathway of photosynthesis reduce CO_2 to phosphoglycerate, a 3-C compound, via the enzyme RuBP carboxylase. This enzyme discriminates against $^{13}CO_2$, resulting in relatively low $\delta^{13}C$ values for C_3 plants. Plants with the C_3 pathway have $\delta^{13}C$ values ranging from approximately -32 to -20 ‰, with a mean of -27 ‰ (Fig. 1).

C_4 plants reduce CO_2 to aspartic or malic acid, both 4-C compounds, via the enzyme PEP carboxylase. This enzyme does not discriminate against ^{13}C as much as RuBP carboxylase does, so that C_4 plants have relatively high $\delta^{13}C$. Values for C_4 plants range from -17 to -9 ‰, with a mean of -13 ‰ (Fig. 1). Thus, C_3 and C_4 species have distinct, nonoverlapping $\delta^{13}C$ values and differ from each other by approximately 14 ‰ (Smith and Epstein, 1971).

Plants with Crassulacean acid metabolism (CAM) are able to minimize water loss by fixing CO_2 at night via the enzyme PEP carboxylase. As a result, most CAM plants have $\delta^{13}C$ values typical of those for C_4 species. However, under certain environmental and developmental circumstances, some CAM species (i.e., facultative CAM species) are able to switch to a C_3 mode of photosynthesis, fixing CO_2 during the day via RuBP carboxylase. These facultative CAM species will have $\delta^{13}C$ values dependent on the relative proportions of carbon fixed by RuBP carboxylase and PEP carboxylase. $\delta^{13}C$ values for CAM plants range from approximately -28 to -10 ‰ but are most commonly -20 to -10 ‰ (Fig. 1).

Most terrestrial plant species are C_3. Most temperate zone and all forest communities are dominated by C_3 species. However, C_4 and CAM plants are significant components of many plant communities, particularly in warm, arid, or semiarid environments (Osmond et al., 1982). For example, tropical and subtropical grasslands consist almost exclusively of C_4 grasses, and CAM plants (e.g., Cactaceae, Euphorbiaceae) are conspicuous in many desert communities. In general, the proportion of C_4 species in a flora increases as latitude and altitude decrease (e.g., Teeri and Stowe, 1976; Boutton et al., 1980).

Although the great majority of organic matter in the terrestrial environment is present as soil organic matter, its isotopic composition is not well known. Certain biochemical fractions of plant litter, the primary input into the soil organic matter compartment, are isotopically distinct and decompose at different rates than other chemical fractions. For example, cellulose and hemicellulose are often 1 to 2 ‰ more enriched in ^{13}C, while lignin is 2 to 6 ‰ lower in ^{13}C than whole plant tissue (Benner et al., 1987). Since lignin decomposes more slowly than cellulose fractions, the opportunity exists for the isotopic composition of litter to change during decomposition. Despite this, the isotopic composition of soil organic matter largely reflects the photosynthetic pathway type of the dominant species in the plant community (Fig. 1). In well-drained mineral soils, organic matter becomes slightly more enriched (1–3 ‰) in ^{13}C relative to the source material with increasing depth in the soil profile (Stout *et al.*, 1981). Because radiocarbon age of soil organic matter increases with depth in the profile, it has been suggested that microbial metabolism during decomposition of the organic matter as it is transported down through the soil profile may be responsible for the observed increase in ^{13}C with depth. The $\delta^{13}C$ value of soil organic matter may not reflect that of the aboveground plant biomass if there has been a recent conversion from one photosynthetic pathway type to another (Dzurec *et al.*, 1985).

Soil CO_2 is derived from root respiration, decomposition of soil organic matter, and, at least in the upper 30 cm of the profile, diffusion of atmospheric CO_2 into the soil. Soil CO_2 is generally about 5 ‰ more enriched in ^{13}C than the associated vegetation and soil organic matter (Fig. 1) possibly because the diffusion coefficients for $^{12}CO_2$ and $^{13}CO_2$ in air differ, allowing $^{12}CO_2$ to diffuse out of the soil more rapidly than $^{13}CO_2$ (Cerling 1984; Cerling *et al.*, 1989; Quade *et al.*, 1989). In the upper portions of a soil profile and in soils where respiration and decomposition are slow due to temperature or moisture limitations, diffusion or relatively enriched atmospheric CO_2 ($\delta^{13}C = -7.8$ ‰) into the soil may cause additional ^{13}C enrichment of the soil CO_2. Thus, the isotopic composition of soil CO_2 will depend on the relative proportions of C_3 and C_4 plants in the extant community (Schonwitz *et al.*, 1986), on the relative proportions of soil organic matter derived from C_3 versus C_4 sources, on diffusive fractionation of $^{13}CO_2$ and $^{12}CO_2$, and on the intensity of soil respiration (Cerling, 1984; Quade *et al.*, 1989).

Where conditions permit the formation of soil carbonates (alkaline, arid or semiarid soils), their isotopic composition is determined by that of the soil CO_2 and by a +10 ‰ equilibrium fractionation during formation of $CaCO_3$ from soil CO_2 (Cerling, 1984). In a pure C_3 plant community, pedogenic carbonates should be near −12 ‰ (−27 + 5 + 10 ‰), while in a pure C_4 community they should be near +2 ‰ (−13 + 5 + 10 ‰) (Fig. 1). Cerling *et al.* (1989) have demonstrated that modern soil carbonates differ systematically from coexisting soil organic matter by 14 to 16 ‰ in undisturbed soils from humid regions. The isotopic properties of soils beneath CAM plant communities have not been investigated but should be related to the isotopic composition of the dominant vegetation.

Stable carbon isotope ratios of dissolved inorganic carbon (CO_2, HCO_3^-, $CO_3^=$) in groundwater at the water table depend on the $\delta^{13}C$ of soil CO_2 and on the $\delta^{13}C$ of dissolved carbonate that may originate from parent materials such as limestone, which is relatively enriched in ^{13}C at 0 ‰ (Fritz *et al.*, 1978). Thus, groundwaters could range from approximately −30 to 0 ‰ but are most often between −25 and −10 ‰ (Fig. 1).

IV. MARINE ENVIRONMENT

By far the largest active pool in the global carbon cycle is the dissolved inorganic carbon (Σ DIC) present in the oceans. This pool is largely HCO_3^- (~95% of carbon in Σ DIC), but also includes dissolved CO_2 (<1% of carbon) and $CO_3^=$ (~5% of carbon). The oceanic Σ DIC pool controls the $^{13}C/^{12}C$ ratio of atmospheric CO_2 and marine carbonates. Both Σ DIC and $CaCO_3$ have mean $\delta^{13}C$ values near 0 ‰ (Anderson and Arthur, 1983; Galimov, 1985), reflecting the fact that there is little fractionation between carbonate ion and $CaCO_3$, and that the PDB standard was a carbonate of marine origin. Within the ocean, photosynthesis and decomposition have the greatest effect on the $\delta^{13}C$ value of Σ DIC. Because the photosynthetic process discriminates against ^{13}C, the residual Σ DIC in surface waters is relatively ^{13}C enriched and has $\delta^{13}C$ values ranging from approximately $+1$ to $+3$ ‰ (Anderson and Arthur, 1983). In addition to photosynthetic effects, the $\delta^{13}C$ of surface water Σ DIC is influenced to a lesser extent by upwelling of deeper water and temperature. In deeper waters, decomposition of ^{13}C-depleted organic matter results in lower $\delta^{13}C$ values of Σ DIC, with values typically near 0 ‰ (Anderson and Arthur, 1983).

Photosynthesis in the marine environment occurs via the C_3 pathway. However, $\delta^{13}C$ values of photosynthetic organisms in the ocean do not always resemble $\delta^{13}C$ values of terrestrial C_3 plants (Fig. 2). This might be related to the use of bicarbonate as a carbon source for photosynthesis, which is substantially higher in ^{13}C than atmospheric CO_2. In addition, slower diffusion of CO_2 in water may reduce the extent of fractionation by RuBP carboxylase (O'Leary, 1988). Other factors that may influence $\delta^{13}C$ values of marine photosynthetic organisms include salinity, temperature, and CO_2 availability.

Phytoplankton have $\delta^{13}C$ values ranging from -30 to -18 ‰ (Degens, 1969; Deines, 1980; Rau *et al.*, 1982; Anderson and Arthur, 1983; Descolas-Gros and Fontugne, 1990) but they are typically near -22 ‰ (Fig. 2). Phytoplankton from high latitudes have slightly lower $\delta^{13}C$ values near -27 ‰ . Macroalgae are substantially higher in ^{13}C, averaging -15 ‰. $\delta^{13}C$ values of -20 ‰ for zooplankton and -17 ‰ for marine vertebrates suggest that there is progressive ^{13}C enrichment along marine food chains. This topic will be addressed in greater detail in Chapter 13, this volume.

Dissolved organic carbon (DOC) consists primarily of soluble products of plankton decomposition and is isotopically similar (Fig. 2) to phytoplankton at approximately -23 ‰ (Williams and Gordon, 1970; Anderson and Arthur, 1983). However, early methods of preparing DOC for $\delta^{13}C$ analysis involved incomplete oxidation, and these values may be in question. Organic

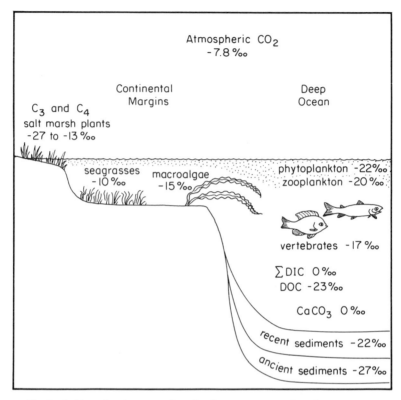

Fig. 2   Stable carbon isotope ratios of major components of marine ecosystems.

carbon in recent and ancient sediments ranges from -20 to -27 ‰ and reflects marine and terrestrial inputs (Degens, 1969; Deines, 1980). Antarctic sediments have slightly lower $\delta^{13}C$ values than those shown in Fig. 2 as a result of lower phytoplankton $\delta^{13}C$ values there.

The marine environment near continental margins is more complicated because carbon sources other than phytoplankton become important. Stable carbon isotope ratios for seagrasses range from -15 to -3 ‰, with an average of -10 ‰ (Fry and Sherr, 1984). Both C_3 and C_4 plant species are represented in salt marshes, with $\delta^{13}C$ values near -27 ‰ for C_3 species and -13 ‰ for C_4 species. $\delta^{13}C$ values for sediments in near-shore environments will therefore depend on the relative proportions of carbon coming from phytoplankton, seagrasses, macroalgae, salt marsh detritus, and carbon transported by rivers. Pollutants are also capable of influencing the isotopic composition of sedimentary organic carbon in the near-shore environment (Calder and Parker, 1968).

Due to the anoxic conditions that prevail in marshes and estuaries, a substantial amount of organic matter in those environments is recycled via bacterial fermentation, resulting in the production of CH_4. In the marine environment, the primary methanogenic pathway is reduction of CO_2, which produces CH_4 with $\delta^{13}C$ values ranging from -110 to -60 ‰ with a mean of -75 ‰ (Whiticar *et al.*, 1986). This loss of ^{13}C-depleted carbon as CH_4 can cause substantial ^{13}C enrichment of Σ DIC in interstitial water in the methanogenic zone, with $\delta^{13}C$ values ranging up to $+15$ ‰ (Whiticar *et al.*, 1986).

V. FRESHWATER ENVIRONMENT

In freshwater lakes and ponds (Fig. 3), the $\delta^{13}C$ of Σ DIC can vary substantially and will depend upon: (1) the extent to which atmospheric CO_2 is in equilibrium with the water mass; (2) seasonal rates of photosynthesis and respiration; (3) the input of CO_2 from decomposition of ^{13}C-depleted terrestrial detritus present in the lake; and (4) the contribution from dissolution of ^{13}C-enriched carbonate rock (Oana and Deevey, 1960; Quay *et al.*, 1986). In general, freshwater systems have lower Σ DIC $\delta^{13}C$ values than marine systems, with values ranging from -15 to 0 ‰. These lower values are due to the presence of ^{13}C-depleted CO_2 derived from decomposition of terrestrial organic matter (Fig. 3). As a result, $CaCO_3$ formed in nonacidic freshwater environments from Σ DIC is more depleted (-15 to 0 ‰) than $CaCO_3$ formed in marine ecosystems. In rivers, Σ DIC ranges from -15 to -5 ‰ (Fry and Sherr, 1984), depending on the relative strengths of organic matter degradation, $CaCO_3$ weathering, and exchange of CO_2 between atmosphere and water.

With the exception of a few species of submerged-aquatic CAM plants, freshwater photosynthetic organisms all appear to have the C_3 pathway. Plankton are generally depleted in ^{13}C and range from -42 to -26 ‰ in lakes (Deines, 1980), and from -30 to -25 ‰ in rivers (Anderson and Arthur, 1983), reflecting the ^{13}C-depleted Σ DIC. Aquatic macrophytes have extremely variable $\delta^{13}C$ values and can range from -50 to -10 ‰, but most are within the range -30 to -12 ‰ (Fig. 3). Reasons for this variability are not well understood but are probably related to the carbon source (HCO_3^- or CO_2), the $\delta^{13}C$ of the carbon source, the slower diffusion rate of CO_2 in water versus air, membrane transport of HCO_3^-, and the presence of an unstirred boundary layer (Smith and Walker, 1980; Osmond *et al.*, 1981). Deviation from the typical C_3 metabolism has been described for several aquatic macrophytes (Bowes and Salvucci, 1989), and may contribute to the isotopic heterogeneity of this group. Lake sediments reflect

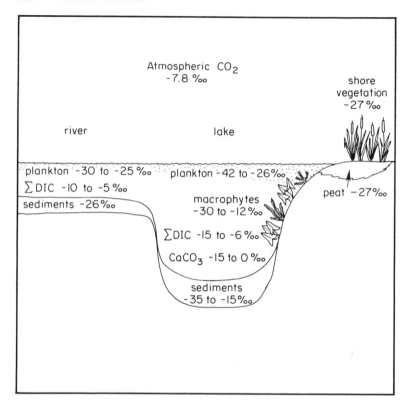

Fig. 3 Stable carbon isotope ratios of major components of freshwater ecosystems.

the isotopic variability of the primary producers, ranging from -35 to -15 ‰ . However, sediments in both lakes and rivers are typically -30 to -26 ‰ (Stuiver, 1975; Hakansson, 1985; B. Fry, personal communication).

Terrestrial vegetation surrounding rivers and lakes will be almost exclusively C_3 plants with $\delta^{13}C$ values near -27 ‰ (Fig. 3). Peat formed in freshwater wetlands usually reflects a C_3 plant source, with $\delta^{13}C$ values of approximately -27 ‰ (Deines, 1980). However, in tropical lowland regions, lakes and rivers may be surrounded by grasses and/or sedges with the C_4 photosynthetic pathway.

Lake sediments and soils associated with freshwater wetlands (swamps, bogs, tundra, etc.) are often anaerobic and organic matter degradation proceeds by fermentation. Thus, these environments constitute a significant source of atmospheric CH_4. In contrast to methanogenesis in the marine environment, CH_4 production in freshwater systems proceeds primarily via acetate fermentation (Whiticar *et al.,* 1986). Methane produced in freshwater wetlands is slightly less depleted in ^{13}C than CH_4 produced in marine environments, with a mean $\delta^{13}C$ value of -60 ‰ and a range from -65 to -50 ‰ (Whiticar *et al.,* 1986). The CO_2 produced as a by-product of methanogenesis via acetate fermentation is relatively ^{13}C enriched, so that Σ DIC in the zone of methanogenesis can become as much as 18 ‰ higher in ^{13}C than Σ DIC outside the zone of methanogenesis (Herczeg, 1988).

VI. CONCLUDING REMARKS

The interesting variations in $\delta^{13}C$ values of natural materials described above can be applied to a wide variety of biological studies, from the physiological to ecosystem level of organization (e.g., Rundel *et al.,* 1989). For example, natural variation in $\delta^{13}C$ can be used to study biochemical and physiological characteristics of photosynthesis (O'Leary, 1988), or aspects of animal nutrition and metabolism (Klein *et al.,* 1986). At the ecosystem level, $\delta^{13}C$ values are useful in following carbon flow through food webs and in identifying sources of carbon contributing to soil organic matter or sediments. Several of these applications are described in more detail in subsequent chapters of this volume.

ACKNOWLEDGMENTS

This manuscript benefited from reviews by David D. Briske, Ethan L. Grossman, and Brian Fry. The assistance of Sylvia Dudash with manuscript preparation is gratefully acknowledged. Support for the preparation of this manuscript was provided by project H-6945 of the Texas Agricultural Experiment Station.

REFERENCES

Anderson, T. F., and Arthur, M. A. (1983). Stable isotopes of oxygen and carbon and their application to sedimentologic and paleoenvironmental problems. In "Stable Isotopes in Sedimentary Geology" (M. A. Arthur, T. F. Anderson, I. R. Kaplan, J. Veizer, and L. S. Land, eds.), pp. 1-1–1-151. Society of Economic Paleontologists and Mineralogists, Tulsa, Oklahoma.

Benner, R., Fogel, M., Sprague, E., and Hodson, R. (1987). Depletion of ^{13}C in lignin and its implications for stable carbon isotope studies. *Nature (London)* **329,** 708–710.

Boutton, T. W., Harrison, A. T., and Smith, B. N. (1980). Distribution of biomass of species differing in photosynthetic pathway along an altitudinal transect in southeastern Wyoming grassland. *Oecologia* **45,** 287–298.

Bowen, R. (1988) "Isotopes in the Earth Sciences." Elsevier Applied Science, New York.

Bowes, G., and Salvucci, M. E. (1989). Plasticity in the photosynthetic carbon metabolism of submersed aquatic macrophytes. *Aquat. Bot.* **34,** 233–266.

Calder, J. A., and Parker, P. L. (1968). Stable carbon isotope ratios as indices of petrochemical pollution of aquatic systems. *Environ. Sci. Technol.* **2**, 535–539.

Cerling, T. E. (1984). The stable isotopic composition of modern soil carbonate and its relationship to climate. *Earth Planet. Sci. Lett.* **71**, 229–240.

Cerling, T. E., Quade, J., Wang, Y., and Bowman, J. R. (1989). Carbon isotopes in soils and paleosols as ecology and paleoecology indicators. *Nature (London)* **341**, 138–139.

Craig, H., Chou, C., Welhan, J., Stevens, C., and Engelkemeir, A. (1988). The isotopic composition of methane in polar ice cores. *Science* **242**, 1535–1539.

Crutzen, P. J. (1983). Atmospheric interactions—Homogeneous gas reactions of C, N, and S containing compounds. In "The Major Biogeochemical Cycles and Their Interactions" (B. Bolin and R. Cook, eds.), pp. 67–112. John Wiley and Sons, New York.

Degens, E. T. (1969). Biogeochemistry of stable carbon isotopes. In "Organic Geochemistry" (G. Eglinton and M. Murphy, eds.), pp. 304–329. Springer, New York.

Deines, P. (1980). The isotopic composition of reduced organic carbon. In "Handbook of Environmental Isotope Geochemistry. Vol. 1, The Terrestrial Environment, Part A" (P. Fritz and J. Fontes, eds.), pp. 329–406. Elsevier Scientific Publishing Company, New York.

Descolas-Gros, C., and Fontugne, M. (1990). Stable carbon isotope fractionation by marine phytoplankton during photosynthesis. *Plant Cell Environ.* **13**, 207–218.

Dzurec, R. S., Boutton, T. W., Caldwell, M. M., and Smith, B. N. (1985). Carbon isotope ratios of soil organic matter and their use in assessing community composition changes in Curlew Valley, Utah. *Oecologia* **66**, 17–24.

Farquhar, G. D., Ehleringer, J. R., and Hubick, K. T. (1989). Carbon isotope discrimination and photosynthesis. *Ann. Rev. Plant Physiol. Plant Mol. Biol.* **40**, 503–537.

Friedli, H., Lotscher, H., Oeschger, H., Siegenthaler, U., and Stauffer, B. (1986). Ice core record of the $^{13}C/^{12}C$ ratio of atmospheric CO_2 in the past two centuries. *Nature (London)* **324**, 237–238.

Fritz, P., Reardon, E., Barker, J., Brown, R., Cherry, J., Killey, R., and McNaughton, D. (1978). The carbon isotope geochemistry of a small groundwater system in northeastern Ontario. *Water Resour. Res.* **14**, 1059–1067.

Fry, B., and Sherr, E. B. (1984). $\delta^{13}C$ measurements as indicators of carbon flow in marine and freshwater ecosystems. *Contrib. Mar. Sci.* **27**, 13–47.

Galimov, E. M. (1985), "The Biological Fractionation of Isotopes." Academic Press, New York.

Golubic, S., Krumbein, W. E., and Schneider, J. (1979). The carbon cycle. In "Biogeochemical Cycling of Mineral-Forming Elements" (P. A. Trudinger and D. J. Swaine, eds.), pp. 29–45. Elsevier Scientific Publishing Co., New York.

Hakansson, S. (1985). A review of various factors influencing the stable carbon isotope ratio of organic lake sediments by the change from glacial to post-glacial environmental conditions. *Quat. Sci. Rev.* **4**, 135–146.

Herczeg, A. L. (1988). Early diagenesis of organic matter in lake sediments: A stable carbon isotope study of pore waters. *Chem. Geol. (Isot. Geosci. Sect.)* **72**, 199–209.

Klein, P. D., Boutton, T. W., Hachey, D. L., Irving, C. S., and Wong, W. W. (1986). The use of stable isotopes in metabolism studies. *J. Anim. Sci.* **63** (Suppl. 2), 102–110.

Levin, I., Kromer, B., Wagenback, D., and Munnich, K. O. (1987). Carbon isotope measurements of atmospheric CO_2 at a coastal station in Antarctica. *Tellus* **39B**, 89–95.

Lowe, D. C., Brenninkmeijer, C. A. M., Manning, M. R., Sparks, R., and Wallace, G. (1988). Radiocarbon determination of atmospheric methane at Baring Head, New Zealand. *Nature (London)* **332**, 522–525.

Mook, W. G. (1986). ^{13}C in atmospheric CO_2. *Netherlands J. Sea Res.* **20**, 211–223.

Mook, W. G., Koopmans, M., Carter, A. F., and Keeling, C. D. (1983). Seasonal, latitudinal, and secular variations in the abundance and isotopic ratios of atmospheric carbon dioxide. 1. Results from land stations. *J. Geophys. Res.* **88**, 10915–10933.

Oana, S., and Deevey, E. S. (1960). Carbon 13 in lake waters, and its possible bearing on paleolimnology. *Am. J. Sci.* **258A**, 253–272.

O'Leary, M. H. (1981). Carbon isotope fractionation in plants. *Phytochemistry* **20**, 553–567.

O'Leary, M. H. (1988). Carbon isotopes in photosynthesis. *BioScience* **38**, 328–336.

Osmond, C. B., Valaane, N., Haslam, S., Uotila, P., and Roksandic, Z. (1981). Comparisons of $\delta^{13}C$ values in leaves of aquatic macrophytes from different habitats in Britain and Finland; some implications for photosynthetic processes in aquatic plants. *Oecologia* **50**, 117–124.

Osmond, C. B., Winter, K., and Ziegler, H. (1982). Functional significance of different pathways of CO_2 fixation in photosynthesis. In "Physiological Plant Ecology II. Water Relations and Carbon Assimilation" (O. L. Lange, P. S. Nobel, C. B. Osmond, and H. Ziegler, eds.), pp. 479–547. Springer-Verlag, New York.

Quade, J., Cerling, T. E., and Bowman, J. (1989). Systematic variations in the carbon and oxygen isotopic composition of pedogenic carbonate along elevation transects in the southern Great Basin, United States. *Geol. Soc. Am. Bull.* **101**, 464–475.

Quay, P. D., Emerson, S. R., Quay, B. M., and Devol, A. H. (1986). The carbon cycle for Lake Washington. A stable isotope study. *Limnol. Oceanogr.* **31**, 596–611.

Rau, G. H., Sweeney, R. E., and Kaplan, I. R. (1982). Plankton $^{13}C:^{12}C$ ratio changes with latitude: Differences between northern and southern oceans. *Deep-Sea Res.* **29**, 1035–1039.

Rundel, P. W., Ehleringer, J. R., and Nagy, K. A. (1989). "Stable Isotopes in Ecological Research." Ecological Studies 68. Springer-Verlag, New York.

Schidlowski, M., Hayes, J. M., and Kaplan, I. R. (1983). Isotopic inferences of ancient biochemistries: Carbon, sulfur, hydrogen, and nitrogen. In "Earth's Earliest Biosphere: Its Origin and Evolution" (J. W. Schopf, ed.), pp. 149–186. Princeton Univ. Press, Princeton, New Jersey.

Schonwitz, R., Stichler, W., and Ziegler, H. (1986). $\delta^{13}C$ values of CO_2 from soil respiration on sites with crops of C_3 and C_4 type of photosynthesis. *Oecologia* **69**, 305–308.

Smith, B. N., and Epstein, S. (1971). Two categories of $^{13}C/^{12}C$ ratios for higher plants. *Plant Physiol.* **47**, 380–384.

Smith, F., and Walker, N. (1980). Photosynthesis by aquatic plants: Effects of unstirred layers in relation to assimilation of CO_2 and HCO_3 and to carbon isotopic discrimination. *New Phytol.* **86**, 245–259.

Stevens, C. M., Krout, L., Walling, D., Venters, A., Engelkemeir, A., and Ross, L. E. (1972). The isotopic composition of atmospheric carbon monoxide. *Earth Planet. Sci. Lett.* **16**, 147–165.

Stout, J. D., Goh, K. M., and Rafter, T. A. (1981). Chemistry and turnover of naturally occurring resistant organic compounds in soil. In "Soil Biochemistry, Vol. 5" (E. A. Paul and J. N. Ladd, eds.), pp. 1–73. Marcel Dekker, Inc., New York.

Stuiver, M. (1975). Climate versus changes in ^{13}C content of the organic component of lake sediments during the Late Quaternary. *Quat. Res.* **5**, 251–262.

Tans, P. (1981). $^{13}C/^{12}C$ of industrial CO_2. In "Carbon Cycle Modelling" (B. Bolin, ed.), pp. 127–129. John Wiley and Sons, New York.

Teeri, J. A., and Stowe, L. G. (1976). Climatic patterns and the distribution of C_4 grasses in North America. *Oecologia* **23**, 1–12.

Whiticar, M. J., Faber, E., and Schoell, M. (1986). Biogenic methane formation in marine and freshwater environments: CO_2 reduction vs. acetate fermentation — Isotope evidence. *Geochim. Cosmochim. Acta* **50**, 693–709.

Williams, P. M., and Gordon, L. I. (1970). Carbon-13 : carbon-12 ratios in dissolved and particulate organic matter in the sea. *Deep-Sea Res.* **17**, 19–27.

12

$^{13}C/^{12}C$ Fractionation and Its Utility in Terrestrial Plant Studies

James R. Ehleringer
Department of Biology
University of Utah
Salt Lake City, Utah 84112

I. INTRODUCTION

A. Variations in ^{13}C Content among Plants

Variation in carbon isotopic composition among plant species was first noted by geochemists in the 1950s. By the 1970s, it was well accepted that the isotopic composition of plants differed widely and that stable isotope ratios could be used to distinguish among photosynthetic pathway types (O'Leary, 1981). Since that time, the use of carbon isotope ratios in plant ecological and physiological research has increased significantly and in several directions (Rundel *et al.*, 1988; Farquhar *et al.*, 1989). Carbon isotope ratios have proven useful as a screen to determine the photosynthetic pathway of a species (C_3 vs C_4), since there are differences in the isotopic discrimination by the initial carboxylating enzyme of each photosynthetic pathway (RuBP carboxylase as the initial carboxylating enzyme in C_3 plants, and PEP carboxylase in C_4 plants). Additionally, the variation in isotopic values at both the interspecific (within a photosynthetic pathway) and intraspecific levels is proving useful in understanding metabolic, environmental, and life history patterns among plants. For example, some aspects of stomatal control in photosynthesis can be evaluated using carbon isotope ratios. Water stress causes a decrease in stomatal activity and, correspondingly, Guy *et al.* (1980) have demonstrated that leaf carbon isotope ratios increase linearly with a decrease in leaf water potential. In a second example, Farquhar and Richards (1984) have shown that carbon isotope ratios can be a reliable indicator of water-use efficiency. Using this as a screen, they have further demonstrated substantial variation in carbon isotope ratio among different genetic lines of wheat, and this variation can be used to select lines that are more water-use efficient in their biomass production. As a last example of how carbon isotope ratios are being applied to botanical studies, Ehleringer and Cooper (1988) have recently shown that there is carbon isotopic variation among species within a plant community and that this variation is tightly associated with life history patterns, with short-lived perennials being less water-use efficient than long-lived components of the community.

B. Notation

Although historically it is most common to express the carbon isotopic composition of plant tissues using the δ notation (see Chapter 10, this volume), there are additional expressions that shed more insight into the mechanisms associated with biological fractionation (see Table 1). Farquhar and Richards (1984) have proposed a more direct measure, Δ, the actual discrimination by the leaf. This expression measures isotopic composition relative to the source of carbon (atmospheric CO_2) rather than to an arbi-

Table 1
Parameters, Symbols, and Units Presented in this Chapter

Symbol	Parameter	Unit
A	Net photosynthetic rate	$\mu mol/m^2/s$
a	Carbon isotope discrimination due to diffusion of CO_2 through the stomatal pore	‰
b	Carbon isotope discrimination due to net C_3 fixation with respect to c_i	‰
b_3	Carbon isotope discrimination due to fixation of gaseous CO_2 by RuBP carboxylase	‰
b_4	Carbon isotope discrimination due to fixation of gaseous CO_2 by PEP carboxylase	‰
c_a	Atmospheric CO_2 concentration	$\mu mol/mol$
c_i	Intercellular CO_2 concentration	$\mu mol/mol$
E	Transpiration rate	$mmol/m^2/s$
g	Leaf conductance to water vapor through the stomata	$mol/m^2/s$
R	Ratio of $^{13}C/^{12}C$	unitless
v	Leaf-to-air water vapor pressure gradient corrected for total atmospheric pressure	$mmol/mol$
Δ	Carbon isotope discrimination by the leaf	‰
α	Isotope effect	unitless
$\varnothing$	Fraction of CO_2 leaking out of bundle sheath cells	unitless
δ	Carbon isotope composition relative to the PDB standard	‰

trary standard such as PDB. The isotope effect (α) associated with carbon accumulation by the leaf is defined as

$$\alpha = \frac{R_{source}}{R_{product}} = \frac{R_a}{R_p} \tag{1}$$

where R_a and R_p denote the $^{13}C/^{12}C$ composition of air and plant, respectively. The discrimination or fractionation by the plant (Δ) is the deviation from unity,

$$\Delta = \alpha - 1. \tag{2}$$

When isotopic composition is expressed as discrimination, plants show a positive discrimination against ^{13}C. Typically, C_3 photosynthetic pathway plants have a discrimination of approximately 22×10^{-3} or, as it is more commonly expressed, 22 ‰. Note that in terms of isotopic composition, δ, relative to the PDB standard,

$$\Delta = \frac{\delta_a - \delta_p}{1 + \delta_p} \tag{3}$$

when Δ is expressed on a per mil basis and where a and p denote air and plant, respectively, making it relatively easy to switch between δ and Δ if the isotopic composition of the source air is known. Note that in Eq. (3) and in Eq. (4), which follows, δ is expressed in absolute units (e.g., use 0.022 or 22×10^{-3}, not 22) (Farquhar *et al.*, 1989). O'Leary (1981) pointed out that the simultaneous use of discrimination and δ is confusing for work with plants, since the discrimination values are usually positive while those of δ are negative. Where possible, it is preferable to use molar abundance ratios ($^{13}C/^{12}C$) only as intermediates in the calculation of final isotope effects. The advantage of reporting Δ values is that it directly expresses the consequences of biological processes, whereas isotope ratio, $\delta^{13}C$, is the result of both source isotopic composition and biological activities.

C. Basis of Isotopic Variation in Plants

Before going out and measuring the isotopic composition of plant materials, it is instructive to have some insight into the mechanistic causes of variation in isotopic composition. The largest differences occur between plants with different photosynthetic pathways. Plants possessing C_3 and C_4 photosynthetic pathways typically differ by approximately 14 ‰ on δ scale (-27 ‰ vs -13 ‰, respectively), which is also equivalent to a differential of approximately 14 ‰ on the Δ scale.

In C_3 photosynthetic plants, the major components of the overall fractionation are the differential diffusion rates of CO_2 through stomata and the fractionation by ribulose bisphosphate (RuBP) carboxylase, the initial carboxylating enzyme of C_3 photosynthesis. Discrimination in C_3 leaves (Farquhar *et al.,* 1982; Farquhar *et al.,* 1989) can be expressed as

$$\Delta = a + (b - a)(c_i/c_a) \qquad (4)$$

where a is the fractionation due to diffusion in air (4.4 ‰), b is the net fractionation caused by carboxylation (27 ‰) (a combination of both RuBP carboxylase and PEP carboxylase activities), and c_i and c_a are the intercellular and ambient partial pressures of CO_2. If stomata are relatively closed, then c_i tends toward zero and Δ approaches 4.4 ‰. On the other hand, if stomatal limitations are minimized, then c_i tends toward c_a and Δ approaches 27 ‰. From studies thus far, c_i/c_a in C_3 leaves appears to range between 0.4–0.8 (Δ range of 13–22 ‰).

The potential advantage of carbon isotope composition studies over instantaneous observations is the long-term integration provided by isotopic analyses. That is, the stable isotopic composition will better reflect the average physiological activity of the leaf over its lifetime than will instantaneous observations of metabolic activity. Additionally, it is possible to sample many plants simultaneously for isotopic composition, allowing for a degree of replication that is simply not feasible using standard gas exchange methodologies.

Carbon isotopic discrimination in C_3 plants provides a long-term indication of c_i/c_a. Additionally, Δ is tightly correlated with water-use efficiency (molar ratio of photosynthesis to transpiration), allowing comparisons of different genotypes or species if $\delta^{13}C_{air}$ has been measured and if leaf temperatures can be considered to have remained the same (Farquhar and Richards, 1984). This follows because photosynthesis (A) can be described as

$$A = (c_a - c_i) \times g/1.6 \qquad (5)$$

where g is the leaf conductance to water vapor transfer and 1.6 is the ratio of diffusivities of water and CO_2 in air. Similarly, transpiration rate (E) can be described as

$$E = v \times g \qquad (6)$$

where v is the leaf-to-air water vapor gradient. Thus, the water-use efficiency (A/E) is

$$A/E = (c_a - c_i)/(1.6 \times v). \qquad (7)$$

Essentially the only variable in this equation is c_i, since c_a and v would be constant for plants growing in the same habitat and with equivalent leaf temperatures.

Discrimination in C_4 photosynthetic pathway plants is slightly more complicated because HCO_3^- (not CO_2) is initially fixed by phosphoenol pyruvate (PEP) carboxylase in mesophyll cells, transported to the bundle sheath cell, and then decarboxylated and recarboxylated by RuBP carboxylase. As discussed by Farquhar (1983), discrimination in C_4 plants can be described by

$$\Delta = a + (b_4 + b_3 \times \varnothing - a) \times c_i/c_a \qquad (8)$$

where b_3 is the discrimination by RuBP carboxylase (29 ‰), b_4 is the discrimination by PEP carboxylase against bicarbonate (-5 ‰), and $\varnothing$ is the leakage rate of CO_2 out of the bundle sheath cells (thought to range between 0.1 and 0.6).

In Crassulacean acid metabolism (CAM) plants, the extent of discrimination will depend on whether CO_2 uptake is occurring only at night (fixation as C_4 acid) or during both the day and night (as in many facultative CAM plants) (Farquhar *et al.*, 1989). In obligate CAM plants, isotopic composition should be similar to that observed in C_4 plants. However, the isotopic composition of facultative CAM plants may span the entire range between C_3 and C_4 values.

II. APPROACHES AND METHODS — SAMPLE COLLECTION

A. Sampling Protocol

Within a plant species, variations in carbon isotopic composition can be the result of source effects, differences in biochemical composition, environmental effects, genetic differences, and/or phenological conditions (O'Leary, 1981). The magnitude of these components may be difficult to anticipate in advance. Thus, when designing an experimental protocol for carbon isotope ratio analyses of plant tissues, it is critical to have a sampling scheme that takes into consideration the potential sources of variation as well as the resolution limitations of the mass spectrometer and preparation techniques (discussed earlier in Chapter 10, this volume). To a large extent, the approach chosen will depend on the purposes behind data collection. For instance, in a broad survey of plant species for photosynthetic pathway determination, few replicate samples are needed and the range in values could be large (e.g., up to 5 ‰), without affecting the interpretation, since we are only interested in classifying plants into one of three broad groupings.

For the most part, sample collection is straightforward. Care should be taken to collect samples at the same time of the day, because diurnal changes in starch and sugar contents can affect carbon isotope ratios. For many studies, collecting at the end of the day is best, because leaf starch and sugar contents are at their peak concentrations. To further minimize sample variation in population-level studies, samples should be collected from the same canopy position. There can be $\delta^{13}C_{air}$ gradients that could potentially confound interpretation of plant isotopic composition values, especially in dense canopies (Farquhar *et al.*, 1989), and care should be made to measure $\delta^{13}C_{air}$ in canopy situations and to correct for this in the data analysis [see air sampling below and recall Eq. (3) and (4)]. After harvesting, the plant sample should be dried immediately. Freezing the tissue until it can be dried or immediate freeze drying of the sample is the recommended procedure. Once dry, the sample can be stored for long periods of time without affecting its isotopic composition. If that is not possible, dry the sample as quickly as possible at a moderate temperature (70–80°C) to avoid loss of organic materials. Few special precautions are necessary for preparing and storing plant samples for later determination of their isotopic composition (i.e., samples can be stored at room temperature for extended periods after drying). Potential changes in chemical composition that accompany long storage and/or slow drying of living material should be avoided.

Consistency in the choice of plant organ to be analyzed is also quite important. While leaves have been commonly used for most measurements in the past, there is now increasing interest in the isotopic composition of other tissues and plant parts (Farquhar *et al.*, 1989). Since the biochemical composition of the tissues or organs often differs from that of the leaves (O'Leary, 1981; Farquhar *et al.*, 1989), it should not be surprising that the isotopic composition may differ as well, thus requiring that there be consistency in the tissues sampled in any sampling scheme.

The isotope ratios of organic material are determined on dried tissues that have been ground to pass through at least a 40-mesh screen. Only small amounts of tissue are required for a carbon isotope ratio analysis. In most cases, less than 3 mg of dried organic material is used for $^{13}C/^{12}C$ determinations. With a cold finger option on the mass spectrometer, less than 0.1 mg can be used in the analysis. With such small samples, the necessity for sample homogeneity cannot be overemphasized. Often there will be a greater variation in the repeated analysis of the same "bulk sample" than in repeated analyses of an individual sample through the mass spectrometer. This is partly because the amount of tissue required for an analysis is usually quite small. Grinding ensures that the sample is homogeneous, and minimizes variation in isotope composition that might exist within the tissue or in any bulked sample. Finely ground material also burns more uniformly during combustion.

B. Variability within a Sample

Sampling protocol and sample sizes should be based on an understanding of the known sources of variation. This will depend, of course, on the tissues or component being analyzed. Possible sources of variation to consider include sample combustion and analysis (usually small relative to other sources of variation) and tissue or biochemical heterogeneity within the sample. Heterogeneity in ground leaves arises because of incomplete grinding and mixing (avoidable) and small differences in metabolic activity or biochemical composition among different leaves in a bulked sample (unavoidable). Genotypic differences between plants may also contribute to this variance (Farquhar and Richards, 1984; Rundel *et al.*, 1988; Farquhar *et al.*, 1989), but as of yet little information is available to assess the possible importance of genetic differences at the population level.

The $^{13}C/^{12}C$ composition of different plant parts is not constant (O'Leary, 1981). There are secondary fractionations in different biosynthetic pathways, and depending on the biochemical composition of that plant part, these may result in a different $^{13}C/^{12}C$ composition than that measured in leaves. It is known that starches and sugars tend to have $^{13}C/^{12}C$ ratios similar to that of the carbon initially fixed in photosynthesis, whereas cellulose tends to be heavier by approximately 2 ‰ and lipids lighter by up to 15 ‰. Thus, in comparing leaves to wood or seed tissues, similar $^{13}C/^{12}C$ ratios should not be expected since the plant parts differ in composition.

To minimize variance in tissue composition, samples can be bulked and well mixed if only a single isotope ratio analysis is feasible. A more productive, but more costly approach, is repeated sampling. In general, the variation in repeated analyses of the same homogeneous leaf fraction (such as starch, sugars, etc.) should have a standard deviation of approximately ±0.05 ‰. Repeated analyses of the same whole leaf tissues should exhibit a standard deviation of approximately ±0.2 ‰. Finally, repeated analyses of leaves at the intrapopulation level (across plants and possible leaf age effects) may exhibit a standard deviation of approximately ±0.4 ‰. Thus, obtaining a 95% confidence interval of ±0.3 ‰ for a population of leaves may requires as many as six independent subsamples.

C. Air Sampling

The $^{13}C/^{12}C$ composition of plant materials will be directly influenced by the isotopic composition of the source material (i.e., atmospheric air). That is, isotopic discrimination by the leaf is added to the isotopic value of the source. Thus, it is absolutely critical to know the carbon isotope ratio of the atmospheric air during the period in which the plant materials were formed. Vastly different results can be obtained when growing plants under natural field versus growth chamber conditions. This is because under many growth chamber conditions the atmospheric CO_2 is affected by human activity (respired CO_2 of approximately -25 ‰), recycled plant respiration (respired CO_2 of approximately -25 ‰), and possibly supplemental CO_2 from bottled tanks (approximate value of -35 ‰). To be on the safe side, it is best to insure a rapid turnover of air with the growth chamber or greenhouse and to insure that all air is drawn from outside of the building.

For field-grown materials, the $^{13}C/^{12}C$ composition of the air is becoming progressively lighter as fossil fuels ($\delta^{13}C_{PDB} \approx -28$ ‰) are added to the atmosphere. Fortunately, the seasonal fluctuations in atmospheric CO_2 concentration (≈ 10 ppm) have only a small effect (<0.2 ‰) on $\delta^{13}C_{air}$, and the relationships between $\delta^{13}C_{air}$ and CO_2 concentration are reasonably predictable (Mook *et al.*, 1983). Under present atmospheric CO_2 concentrations of 345 ppm, the $\delta^{13}C_{air}$ is approximately -7.8 ‰. With well-mixed atmospheric conditions, $\delta^{13}C_{air}$ need be measured only a few times during the growing season. There are, however, conditions in which $\delta^{13}C_{air}$ will need to be measured more frequently. Such conditions exist whenever the CO_2 concentration is affected by decomposition or respiration processes and in urban areas (Keeling, 1958). In extreme cases, CO_2 concentrations increase as much as 100 ppm because of decomposition and nighttime respiration on a tropical rainforest floor or because of automobile or factory exhaust in cities.

Under many growth chamber and greenhouse conditions, bottled CO_2 is used to maintain atmospheric CO_2 levels at a constant value. Unfortunately, the $\delta^{13}C_{air}$ of bottled air is approximately -35 ‰ (personal observation), which is quite different than that of atmospheric CO_2 (≈ -8 ‰). Without taking these source differences into consideration, there could be an error of as much as 27 ‰ in the leaf discrimination estimate.

D. Sample Variability

As long as the investigator is aware of the possible sources of variability affecting $^{13}C/^{12}C$, an appropriate sampling scheme can be designed. The most important sources of variability under field conditions that the investigator should be aware of are *source air* and *environmental heterogeneity*. Changes in source CO_2, as described above, are usually not large, but can be important when bottled versus ambient air is used or whenever respiration and decomposition processes are increasing CO_2 concentrations. Changes in plant carbon isotope ratios due to environmental heterogeneity are most likely associated with either large differences in soil moisture content (affecting plant water status) or light intensity, both of which are parameters known to influence the c_i/c_a ratio. Leaf carbon isotope ratios may be changed by $1-5$ ‰, depending on plant water status and light levels during growth. Seasonality and phenological development, both of which coincide with changes in climatic conditions, should be expected to exert an influence on the carbon isotope ratio through changes in c_i/c_a ratios.

III. PROCEDURE

A. Photosynthetic Pathway Determination

Very small quantities (less than 3 mg dry tissue) and a single sample are all that are generally needed to determine the photosynthetic pathway of a plant. This is perhaps the coarsest of all the procedures, because essentially any part of the plant can be used for analysis. Clear identification of the sample to the species level is perhaps the limiting step. Although not the recommended procedure, in many of the past surveys, samples have been collected from a portion of an herbarium voucher. Insecticides or other sprays to preserve the voucher can affect the isotopic composition.

After sample combustion (described in Chapter 1, this volume), the photosynthetic pathway of the species can be determined based on the following:

	$\delta^{13}C$	Δ
C_3 photosynthesis	−21 to −35 ‰	13–22 ‰
C_4 photosynthesis	−10 to −14 ‰	4–8 ‰
Facultative CAM	−15 to −20 ‰	8–13 ‰
Obligate CAM	−10 to −14 ‰	4–8 ‰

Separating C_4 and CAM plants can be difficult based on $\delta^{13}C$ alone. However, as a general rule, if $\delta^{13}C$ is between −10 and −15 ‰ and the photosynthetic tissue is succulent, then the plant was CAM. On a more definitive basis, C_4 and CAM plants can be distinguished on the basis of differences in their deuterium to hydrogen ratios (Sternberg *et al.*, 1986).

$\delta^{13}C$ ratios in the range of −10 to −20 ‰ have been reported for a number of aquatic plants. It was initially thought that this might indicate the presence of C_4 photosynthesis. However, virtually all higher aquatic plants have C_3 photosynthesis (the exceptions being *Hydrilla, Isoetes,* and *Udotea,* submerged CAM plants). It is now known that CO_2 diffusion limitations can be extreme in the unstirred layers of some aquatic environments, and this is the basis for such high $^{13}C/^{12}C$ ratios (Raven, 1987).

B. Intercellular CO_2 Determination in C_3 Plants

In theory, if the isotopic composition of the source air ($\delta^{13}C_{air}$) is known, then Eq. (3) and (4) indicate that carbon isotopic analyses of plant tissues will then provide an estimate of the c_i/c_a ratio. While entire leaf tissues are used for most carbon isotope ratio analyses, there are occasions for which isotopic analyses are made on leaf and/or tissues fractions. In particular, carbon isotopic composition is frequently measured on sugar, starch, and cellulose fractions. Each has their advantage. Whereas carbon isotopic composition of an entire leaf might best reflect the time-averaged c_i/c_a ratio of a

C_3 leaf over its lifetime, analyses of sugars will more closely reflect the c_i/c_a ratio over the past 1–2 days and starches over the past 4–6 days. Thus, by analyzing sugar or starch we can get a better indication of how plants are responding to particular treatments of interest (such as exposure to an air pollutant, short-term drought stress, or to increased temperatures). On the other hand, long-term analyses (longer than the lifetime of a single leaf) are perhaps best obtained through tree ring analyses. Tree rings can provide a record over ten- to several thousand-year periods. Since tannins and other components can be laid down in the wood of these rings at a later date, only the cellulose fraction (which was laid down at the time of ring formation) should be used for tree ring studies.

C. Sugar Extraction

Fresh leaf material should be killed as soon as possible, preferably by plunging the tissue into liquid nitrogen, in order to minimize any potential for changes in composition. Freeze dry the tissue if possible and grind to at least 40 mesh. The method below is adapted from Brugnoli *et al.* (1988). Place 100 mg sample in a 50-ml centrifuge tube. Add 35 ml distilled water and bring to a boil for 30 min. After cooling, centrifuge at 12,000 g for 15 min. The soluble sugars will be in the supernatant. Collect 30 ml supernatant and mix with 0.3 g DOWEX-50W (pretreated by washing with 1N HCl) for 20 min. Filter with a Buchner funnel through Whatman #1 filter paper. Mix the filtrate with 0.45 g treated DOWEX-1 for 20 min. Again filter with a Buchner funnel through a Whatman #1 filter paper. If trace amounts of chlorophyll are visible, they can be removed from the filtrate: attach a prewetted C18 Sep-Pak cartridge (Water Assoc., Milford, Massachusetts) to a 20-cc Luer-Lock syringe, pour the filtrate into the syringe, and squeeze the solution through the cartridge to remove chlorophyll. Then pour the filtrate into a glass beaker over a low heat until about 1 ml solution remains. Pipet the remaining solution onto a quarter circle of ashed glass filter paper (25-mm diameter) supported by a watch glass. Evaporate the remaining solution on the filter paper in a drying oven at 70°C for 24 hr. The sugar dries firmly on the glass filter paper, which then provides a convenient means of loading small amounts of sample for carbon isotope analysis.

D. Starch Extraction

Starch can decompose if slowly dried, and so it is best to immediately freeze leaf tissue in liquid nitrogen. Freeze dry the tissue if possible and grind to at least 40 mesh. The following procedure is also adapted from Brugnoli *et al.* (1988). Place 100 mg sample in a 50-ml centrifuge tube. Add 35 ml distilled water and bring to a boil for 30 min. After cooling, centrifuge at 12,000 g for 15 min. Resuspend the pellet in 25 ml of 80% boiling ethanol. Then centrifuge at 12,000 g for 10 min. Discard the supernatant. Repeat the resuspension in ethanol and centrifugation until the supernatant is colorless. Resuspend the pellet in 10 ml of 20% hydrochloric acid (w/w) for 30 min. Centrifuge at 12,0000 g for 10 min. Save the supernatant. Resuspend the pellet in 10 ml of 20% hydrochloric acid. Centrifuge as before. Combine the two supernatants and add 25 ml of 80% ethanol. Let this solution stand for 12 hr. Then centrifuge the solution at 12,000 g for 20 min. Discard the supernatant; the pellet represents starch. Resuspend the pellet in 1 ml distilled water and pipet onto a quarter circle of ashed glass filter paper supported by a watch glass. Evaporate the remaining solution on the glass filter paper in a drying oven at 70°C for 24 hr. The starch dries firmly on the glass filter paper, which then provides a convenient means of loading small amounts of sample for carbon isotope analysis.

E. Cellulose Extraction

Cellulose extraction should be carried out under a fume hood. The procedure described below is Sternberg *et al.*'s (1986) modification of the standard extraction method described by Wise (1944). For a final yield of approximately 100 mg, begin with approximately 1 g of dry, ground leaf material or 400 mg of wood tissue. Mix the ground plant material with 200 ml of distilled water in a 400-ml glass beaker. Boil the mixture for 1 hr on a hot plate, constantly stirring the solution and maintaining a volume of 200 ml. After 1 hr, cool the solution to 70°C and maintain this temperature throughout the rest of the extraction procedure. Add 1 ml 100% acetic acid and 1 g sodium chlorite (these materials are toxic, so wear gloves). Stir the solution for 1 hr. Repeat the additions of acetic acid and sodium chlorite at 1-hr intervals four times, for a total of 5 ml acetic acid and 5 g sodium chlorite. Decant the supernatant, wash, and rinse the residue three times with distilled water. Then add 100 ml of 17% sodium hydroxide (w/v) and let stand for 45 min. Decant, wash, and rinse the residue three times with distilled water. Add 100 ml of 10% acetic acid and let stand for 15 min. Decant, wash, and rinse the residue three times with distilled water. Add 200 ml of distilled water to the residue and let stand for 12 hr. Decant water and then place the remaining cellulose – water mixture into a centrifuge tube and centrifuge at 12,000 g for 30 min. Decant the supernatant from the centrifuge tube and place the cellulose pellet on a teflon-coated watch glass to air dry. After drying, the cellulose pellet can be peeled off. Oven dry the cellulose at 50–60°C for 12 hr and then the sample is ready for analysis.

F. On-Line Discrimination Measurement

Carbon isotope discrimination can be measured in a nondestructive manner using gas exchange approaches (Evans *et al.,* 1986). This is a new and powerful approach, with the advantage that changes in isotope discrimination of intact tissues can be measured over short time intervals in response to changes in environmental parameters.

In the conventional gas exchange approach, a leaf or other organ is enclosed in a well-mixed cuvette. Gas exchange processes (photosynthesis, respiration, and transpiration) are measured by the changes in the concentrations of gases before and after the air has passed through the cuvette, taking into consideration also the flow rate and plant surface area. Since photosynthetic processes [Eq. (4) and (8)] discriminate against ^{13}C, the residual CO_2 in air leaving the cuvette should be enriched in ^{13}C relative to that entering the cuvette. The discrimination by the plant tissue (Δ) will be

$$\Delta = \frac{\xi(\delta_o - \delta_e)}{1 + \delta_o - \xi(\delta_o - \delta_e)} \tag{9}$$

where

$$\xi = \frac{c_e}{c_e - c_o}. \tag{10}$$

c_e and c_o are the mole fractions of CO_2 in air, measured at a standard humidity, entering and leaving the cuvette, respectively, and δ_e and δ_o are the carbon isotope ratios of the CO_2 in air entering and leaving the cuvette, respectively. Remember that δ in Eq. (9) is expressed in absolute units (e.g., 0.022 or 22×10^{-3}, and not 22) (Farquhar *et al.*, 1989). The oxygen isotopes in CO_2 that exits the cuvette will also be enriched relative to incoming CO_2. A mass spectrometer with a triple collector should be used to correct for the oxygen isotopic enrichment, since the ^{13}C correction is actually made with ^{17}O calculated from ^{18}O (Santrock *et al.*, 1985). When bulk air from before and after the cuvette is trapped for carbon isotope ratio analysis, N_2O will also freeze out along with CO_2. Since N_2O will significantly affect the calculated $^{13}C/^{12}C$ ratio, it is best when this gas is removed (described in the next section).

G.   Air Sampling

Carbon isotope ratio analysis of atmospheric CO_2 is an extremely important baseline for whole tissue studies as well as being very useful for estimating the fraction of decomposition- and respiration-derived CO_2 in the air. Unfortunately, since carbon dioxide is a trace gas (only 0.035% of the total volume), it is one measure that is all too often ignored by biologists (perhaps because it is thought to remain constant).

Two approaches (variable vs constant volumes) have been used to collect air CO_2 samples for ^{13}C analysis in the field. In the variable volume approach, air is slowly drawn by a pump through condensing traps to first remove water vapor (a $-86°C$ methanol-dry ice slurry), and then through a second trap to freeze out the CO_2 (liquid nitrogen). Both of the traps consist of 6–8 coils with the upper portions of the coil at a nonfreezing temperature. Gas molecules that freeze as aerosols in the moving gas stream, as opposed to those molecules condensing and freezing out on the sides of the tubing, have a tendency not to stick and to continue moving in the air stream. To insure that the gas molecules not caught on the first pass will be trapped later on, multiple loops are used. The upper portion of the loops should be kept at a noncondensing temperature (i.e., $> -100°C$ for CO_2). This general approach can be classified as the variable volume approach, since as pumping times and flow rates vary, the amount of air sampled will differ. There are several major drawbacks to this approach. First, dry ice and liquid nitrogen may be difficult to obtain under many field conditions. Second, traps can get clogged with either dust and/or frozen gases. Last and most important, if flow rates are not sufficiently low and well controlled, not all of the CO_2 will be trapped as it flows through the traps, leading to a fractionation and overestimate of the $^{13}C/^{12}C$ ratio (i.e., heavier values than expected for CO_2 in air).

The constant-volume approach is to draw air with a pump through a glass vessel (usually 1–2 liters). The larger the volume and slower the flow rate, the better time averaged the sample will be. Sample vials should be preevacuated or back-filled with N_2 before the sample is collected in order to insure that CO_2 contamination does not occur. Vacuum-tight stop cocks are used at the inlet and outlet of the collection vial to insure that no gas exchange occurs once the sample has been collected. After the vials have been transported back to the laboratory, CO_2 is extracted from the collection vial by connecting it to a vacuum line, containing first a water vapor trap and second a spiral or loop CO_2 trap. The major drawback of this approach is that relatively large and bulky collection vials are needed. For most studies, this is not a major concern. However, in sampling remote areas, it can restrict the numbers of samples that can be collected.

When using either of the air sampling techniques, the CO_2 sample will not be pure because of contamination by N_2O. Nitrous oxide is a biogenic trace gas in the atmosphere (ratio of N_2O/CO_2 is approximately 0.89×10^{-3}) that condenses out with carbon dioxide at liquid nitrogen temperatures. This is an obvious concern since $^{12}C^{16}O^{16}O$ and $^{14}N^{14}N^{16}O$ have identical masses and are thus indistinguishable in the mass spectrometer. Furthermore, even these trace amounts of N_2O can alter the calculated $^{13}C/^{12}C$ ratio by approximately 0.23 ‰. The N_2O can be removed through combustion. The CO_2–N_2O sample is circulated for 15 Min through copper pellets at 650°C, after which CO_2 is frozen and the N_2 is pumped off (Mook and Jongsma, 1987). The primary drawback to this combustion is that the $\delta^{18}O$ signal in carbon dioxide will be modified by the combustion process. Mook and

Jongsma (1987) point out that it is possible to avoid this combustion process and to correct atmospheric samples by adding 0.23 ‰ to the observed $\delta^{13}C$ value and 0.33 ‰ to the $\delta^{18}O$ value.

REFERENCES

Brugnoli, E., Hubick, K. T., von Caemmerer, S., and Farquhar, G. D. (1988). Correlation between the carbon isotope discrimination in leaf starch and sugars of C_3 plants and the ratio of intercellular and atmospheric partial pressures of carbon dioxide. *Plant Physiol.* **88,** 1418–1424.

Ehleringer, J. R., and Cooper, T. A. (1988). Correlations between carbon isotope ratio and microhabitat in desert plants. *Oecologia* **76,** 562–566.

Evans, J. R., Sharkey, T. D., Berry, J. A., and Farquhar, G. D. (1986). Carbon isotope discrimination measured concurrently with gas exchange to investigate CO_2 diffusion in leaves of higher plants. *Aust. J. Plant Physiol.* **13,** 281–292.

Farquhar, G. D. (1983). On the nature of carbon isotope discrimination in C_4 species. *Aust. J. Plant Physiol.* **10,** 205–226.

Farquhar, G. D., and Richards, R. A. (1984). Isotopic compositoin of plant carbon correlates with water-use efficiency of wheat genotypes. *Aust. J. Plant Physiol.* **11,** 539–552.

Farquhar, G. D., O'Leary, M. H., and Berry, J. A. (1982). On the relationship between carbon isotope discrimination and intercellular carbon dioxide concentration in leaves. *Aust. J. Plant Physiol.* **9,** 121–137.

Farquhar, G. D., Ehleringer, J. R., and Hubick, K. T. (1989). Carbon isotope discrimination and photosynthesis. *Annu. Rev. Plant Physiol. Mol. Biol.* **40,** 503–537.

Guy, R. D., Reid, D. M., and Krouse, H. R. (1980). Shifts in carbon isotope ratios of two C_3 halophytes under natural and artificial conditions. *Oecologia* **44,** 241–247.

Keeling, C. D. (1958). The concentration and isotopic abundances of atmospheric carbon dioxide in rural areas. *Geochim. Cosmochim. Acta* **13,** 322–334.

Mook, W. G., and Jongsma, J. (1987). Measurement of the N_2O correction for $^{13}C/^{12}C$ ratios of atmospheric CO_2 by removal of N_2O. *Tellus* **39B,** 96–99.

Mook, W. G., Koopmans, M., Carter, A. F., and Keeling, C. D. (1983). Seasonal, latitudinal, and secular variations in the abundance of isotopic ratios of atmospheric carbon dioxide. 1. Results from land stations. *J. Geophys. Res.* **88,** 10915–10933.

O'Leary, M. H. (1981). Carbon isotope fractionation in plants. *Phytochemistry* **20,** 553–567.

Rundel, P. W., Ehleringer, J. R., and Nagy, K. A. (eds.) (1988). "Stable Isotopes in Ecological Research." Springer Verlag, New York.

Raven, J. A. (1987). Application of mass spectrometry to biochemical and physiological studies. In "Biochemistry of Plants" (D. D. Davies, ed.), vol. 13, pp. 127–180. Academic Press, New York.

Santrock, J., Studley, S. A., and Hayes, J. M. (1985). Isotopic analyses based on the mass spectrum of carbon dioxide. *Anal. Chem.* **57,** 1444–1448.

Sternberg, L. S. L., DeNiro, M. J., and Johnson, H. B. (1986). Oxygen and hydrogen isotope ratios of water from photosynthetic tissues of CAM and C_3 plants. *Plant Physiol.* **82,** 428–431.

Wise, L. E. (1944). "Wood Chemistry." Reinhold, New York.

13

The Study of Diet and Trophic Relationships through Natural Abundance ^{13}C

Juanita Newman Gearing

Fisheries and Oceans Canada,
Maurice Lamontagne Institute,
Mont-Joli, Quebec, Canada G5H 3Z4

I. INTRODUCTION

Measurement of the natural abundances of stable carbon atoms ($\delta^{13}C$) is a powerful way to study the dietary organization of an entire ecosystem. In some cases, it is often the only tool available for distinguishing and tracing different food sources. However, like all methods, it has certain limitations; care must always be taken in designing the experiment and in interpreting the results. This chapter will attempt to cover all of these aspects of the method. Section I gives examples of specific areas in which the technique is valuable, a general discussion of its limitations and how to work around them, and finally some examples of particular studies that are used to illustrate the different procedures. Section II enumerates the various factors that may introduce uncertainty into the final results; these parameters are important for both experimental design and for interpretation of results. Section III describes the particular steps necessary to obtain, process, and preserve samples. General techniques and principles are given that can be applied to all ecosystems. Specific methods for marine plankton illustrate how the principles can be applied. Procedures for combustion and mass spectrometric analyses are found in Chapter 10, this volume.

A. Importance

Natural-abundance isotope ratios are one of the few techniques for the examination of carbon pathways in the field and throughout entire, natu-

Table 1
Comparison of Two Methods for Determination of Diet

Gut contents	$\delta^{13}C$
Immediate food	Average food (over weeks or months)
Ingested food	Assimilated food
Particular food	General type of food
Unambiguous affirmation of food ingestion	Unambiguous denial of food usage

rally functioning ecosystems. In nature, enrichment studies with either stable or radioactive compounds, while being extremely sensitive, are usually too expensive and environmentally undesirable. Stable isotope ratios make use of the small, natural differences to provide quick, cost-effective surveys of entire ecosystems, pointing the direction for future intensive work in either the field or the laboratory.

Isotope ratios are a useful complement to more traditional ways to study diets: direct observations and analyses of gut contents. The methods give quite different information (Table 1). Gut contents are good for determining the particulars of feeding at one point in time. Provided that the food does not break down rapidly, such as happens with many small particles (including marine plankton and larval fish), examination of gut contents can unambiguously ascertain that a specific plant or animal was ingested. Isotope ratios, on the other hand, are determined by the general type of food (original method of carbon fixation, number of trophic levels) that has been incorporated into the animal over the past several weeks or months. It gives a good overall idea of the average diet. With multiple types of food generally available, isotope ratios can indicate but cannot prove that a certain type of food was used; they can, however, sometimes prove when a food has *not* been eaten and assimilated.

B. Limitations

Understanding the limitations of a method is of paramount importance for using it to the fullest extent and for minimizing the standard error. For work with natural abundances the most obvious limitation, as well as the greatest advantage, is that nature has provided the label at a low level in the field; no extrapolation to nature from artificial environments is necessary. In the laboratory, enrichment studies offer much more sensitivity. Natural abundances, however, are ideally suited to studies in many natural environments. Trophic level studies can be done where the possible carbon sources have approximately the same isotopic signal or when the carbon source is known. These types of studies are best done, however, in conjunction with ^{15}N ratios since the latter have a larger change per trophic level than ^{13}C — approximately 3 ‰ versus 1 ‰ (DeNiro and Epstein, 1978, 1981).

Tracing one particular type of carbon requires that it be isotopically distinct. Surveys of the isotope ratios of different carbon sources can be found in Chapter 10, this volume, and in reviews such as van der Marwe (1982), Rounick and Winterbourn (1986), Peterson and Fry (1987), and Gearing (1988). The larger the naturally occurring isotopic difference, the more sensitive and precise the technique. The smaller the isotopic difference, the more samples are necessary to see statistically significant differences. For example, differences of about 5 ‰ can be distinguished with only a few samples, 2–3 ‰ changes may require ten or more samples, and 1 ‰ changes can be significant with several tens of samples although the standard error is usually great.

Some environments are quite complex, having several isotopically different organic sources. For example, much work has been done on seagrass and saltmarsh ecosystems in order to determine the relative importance of land plants, marsh macrophytes, and microalgae in these highly productive ecosystems. In such places, studies using multiple isotopes (^{15}N, ^{34}S, ^{14}C, etc.) have been useful for sorting out the flow of organic matter (e.g., Peterson and Howarth, 1987).

No matter how simple or complex the environment, there is always the possibility of an unknown or previously unconsidered carbon source, or isotopic deviations in a known source at a particular locale. For example, isotopic evidence indicated the importance of small, little-considered epiphytic algae as a carbon source in some seagrass meadows (Kitting *et al.,* 1984). Spatial isotopic changes are due to such factors as climate, light, and species (see Chapter 12, this volume). Thus it is impossible for isotopic data to irrefutably prove that a certain carbon source was used. Potential errors are minimized by examining as many carbon sources as possible at a particular site during the course of the experiment. If unexpected values are found despite these measurements, one can calculate back to a postulated carbon source (Peterson *et al.,* in preparation). This procedure can point out areas for further research.

Finally, but most importantly, researchers must remember that no single type of data is strong when it stands alone. Even in studies whose principal thrust is isotopic, it is vital to have as much other information (physical, chemical, and biological) as possible, not only for interpreting the isotopic data but also for choosing the appropriate sampling sites, times, and organisms. Complementary independent measurements may substantiate or further constrain interpretation of the isotopic evidence. Later sections of this chapter will elaborate on how the $\delta^{13}C$ of an animal may change, for example, with seasons, with its biochemistry, and with its stage of growth. It is important when designing a sampling program to try to reduce this variability to a minimum (unless measuring the variability in a population is one of the desired results).

C. Uses

The primary theoretical basis of using $\delta^{13}C$ as a tracer is that the characteristic ratios of different sources are preserved as the carbon is cycled through organisms and detritus; that is, you are what you eat. This assumption seems to be, within limits of uncertainty, correct. However, there is a trend to enrichment in ^{13}C with increasing trophic level, about 1 ‰ per level (DeNiro and Epstein, 1978; Gearing *et al.*, 1984, and references therein).

These properties have allowed $\delta^{13}C$ values to be used for three types of problems. A few recent examples of each type are given below; they will be used in later sections to illustrate different methods. More extensive bibliographies of $\delta^{13}C$ work can be found in the papers cited and the reviews mentioned above.

Most of the work to date has used $\delta^{13}C$ for tracing a particular carbon source. Studies can either examine entire ecosystems or focus on animals of the same species or feeding type to avoid variability due to trophic level. Animals and their possible food sources are measured and the relative importance of each food calculated using simple mixing equations (see Section IIIG). Research has included a distinction between terrestrial and freshwater production (Rounick and Hicks, 1985) as well as between terrestrial and marine production (Newman *et al.*, 1973; Incze *et al.*, 1982). Isotopes have been used to trace carbon from salt marshes (Peterson and Howarth, 1987), mangroves (Rodelli *et al.*, 1984), sewage (Oviatt *et al.*, 1987), petroleum (Spies and DesMarais, 1983), chemosynthetic bacteria (Kennicutt *et al.*, 1985), and eutrophic primary production (Gearing *et al.*, 1991). On land they have been used to differentiate grasses from shrubs (Tieszen *et al.*, 1979), and in the sea to distinguish different sizes of phytoplankton (Gearing *et al.*, 1984).

Carbon isotopes may also be used to determine trophic relationships in which one can assume a common food base. For example, McConnaughey and McRoy (1979) and Gearing *et al.* (1984) have examined various estuarine and pelagic marine ecosystems.

Finally, under certain circumstances in which the source of food changes, $\delta^{13}C$ can be used to determine turnover rates in the same manner as carbon-14 or enrichment studies in the laboratory. A few of these types of studies have been carried out in the laboratory (Fry and Arnold, 1982; Tieszen *et al.*, 1983) and in the field (Jones *et al.*, 1981; Peterson *et al.*, in preparation). Natural migrations can also be inferred from isotopic evidence of turnover (Fry, 1981).

II. SOURCES OF VARIABILITY

In order to design a successful isotopic study, the sources of natural variability must be understood. If ignored, ecological isotopic variations could lessen the statistical validity of the results. However, if variability is taken into consideration from the beginning of a study, efficiency can be increased (maximum precision of results from a given number of analyses). In some cases, these sources of variability may even point out areas where the isotopic results give additional insights into the functioning of the ecosystem.

Methods of sample combustion to CO_2 and analysis by mass spectrometry are given in Chapter 10, this volume. The error associated with this laboratory work ("analytical variability") is usually quoted as $\pm 0.1 - 0.3$ ‰, depending on the type and size of sample, combustion method, and type of mass spectrometer used.

For studying natural diets, a second source of error, which can be called "ecological variability," must also be considered. This is summarized in Table 2. Ecological variability is the 0.2 to 2 ‰ standard deviation found for animals of the same species raised in similar environments on the same food (Anderson *et al.*, 1987; DeNiro and Epstein, 1978; Fry, 1981; Teeri and Schoeller, 1979; Tieszen *et al.*, 1983). This variability averages around 0.6 ‰ and should always be measured during the experiment if quantitative results are desired. Good isotopic studies incorporate a sampling design to determine at the outset the isotopic variability at each site.

In order to assess the relative importance of different food sources, it is often of interest to minimize ecological variability in order to obtain the most accurate average $\delta^{13}C$ value for a species or group of animals. Doing

Table 2
Extent of Isotopic Variability

Source	Isotopic change ($\delta^{13}C$, ‰)		
Mass spectrometry	<0.02		
Combustion	<0.3		
Inhomogeneity of food	up to 8		
Change of food[a]	up to 30		
Intraspecific	<2		
	Size	0.5–2	
	Sex	0	
	Health	0.8	
	Age	0.4–2	
Different organs	<3		
Seasonality	1–4		

[a] Depends on the isotopic difference in available foods.

this requires an understanding of its causes. Several different sources of variability have been hypothesized and studied in limited environments. Unfortunately no comprehensive study of their mechanisms and relative importance has yet been published.

A. Inhomogeneity In a Single Food Source

The first possible source of variability is food. Plants with the same photosynthetic pathways, even those of the same species, can differ in $\delta^{13}C$ values depending on growth conditions (Chapter 12, this volume). Stephenson *et al.* (1984) reported a range of up to 8 ‰ within a macroalga; more typical values are 1–2 ‰. Animals raised on isotopically homogeneous food in the laboratory still have an average standard deviation of about 0.6 ‰. The effect of inhomogeneous food is counteracted by the fact that animal tissue represents an average of food assimilated over time. The result seems to be increased variability for a few omnivores but no magnification of variability with trophic position.

B. Selectivity

Isotopic changes may be caused by selective feeding and selective assimilation. These are often the parameters being investigated in the study. Whether they can affect the variability within a species, that is to say, whether individuals differ in their feeding preferences or in their ability to digest foods, is not known. Analyses of more single animals will give a more accurate average, but will not necessarily lower the intraspecific variability. By treating several individuals as the same sample, an accurate average can be obtained with fewer analyses. However, the slightly lower variability of these numbers is not an accurate reflection of environmental variability; the true variability must be used when assessing the standard error associated with the calculation of relative food usage.

C. Biochemical Differences

Because of isotopic fractionation during metabolism, different biochemicals in the same animal have different isotope ratios. Major biochemical fractions such as carbohydrates, proteins, and lipids may differ by 5 to 10 ‰ while differences of 2 to 5 ‰ (or more) can be found between different lipid classes or individual amino acids. Thus individuals of the same species may vary depending on their biochemical state (for example: before and after spawning, healthy or sick, fat or lean). There will also be differences between species that may eat the same food but store different amounts of fat. One study has attempted to correct for the relative amounts of lipids (isotopically more negative) present in different species (McConnaughey and McRoy, 1979). Such differences make it imperative that each specimen be examined to determine as much as possible about its general state (size, sex, health, etc.). Slight differences have been found with size in some species but not in others (Fry and Parker, 1979; Gearing *et al.,* 1984); no differences were found with sex (Fry and Parker, 1979); some disparities were noted between unhealthy individuals (Gearing and Gearing, 1991); and a 0.8 ‰ difference has been related to age (Yoneyama *et al.,* 1983). The extent of and reasons for such variability are important areas of study that have been neglected to date.

Studying different biochemicals isotopically within particular organisms can provide valuable information on metabolic pathways. If, on the other hand, the purpose of the study is to examine the relationships between different organisms (typical feeding and pollution studies), biochemical differences can be minimized by analyzing a sample that is chemically more defined, that is, to isolate a class of chemicals or even individual compounds. Researchers have measured total lipids, hydrocarbon classes, amino acids, and other fractions. Of particular interest is the analysis of bone collagen (see Schoeninger and DeNiro, 1984; Tuross *et al.,* 1988, and references therein) because the organic matter contains both carbon and nitrogen for analysis, and the organic matter is relatively well preserved, allowing the analysis of old as well as recent diets. There are at least two problems associated with analyzing specific chemicals: analysis time and possible fractionation during workup. The extraction and purification procedures required to isolate compounds or chemical classes are time consuming and expensive, although new technology is beginning to lessen the time required. Depending on the substance, isolation may take from a few hours to several days. It also increases the collection time because a larger quantity of tissue is required. Thus, instead of processing several samples a day, a laboratory may be capable of analyzing only one sample every few days (or even weeks in extreme cases). Moreover, each of the procedures used in the isolation must be checked to see if it fractionates the carbon isotopes; 1 to 2 ‰ changes during workup are not uncommon. Thus, the use of individual chemicals for analysis offers many advantages and opens up new areas of study, but it does this at the cost of time, effort, and possible laboratory fractionation.

D. Intraorganism Differences

Distinct parts of the body may differ isotopically (summarized in Table 3). Ranges average about 3 ‰ for both terrestrial and marine animals. These disparities probably arise from differences in biochemistry, and thus may change with time within an individual. Similar differences have been noted in plants; for instance leaves of trees are generally more depleted in ^{13}C than woody tissue (Leavitt and Long, 1982). It is much simpler to isolate an organ

Table 3
Isotope Ratios in Different Animal Organs Compared to that in Muscle[a,b]

$$\Delta = \delta^{13}C_{organ} - \delta^{13}C_{muscle}$$

	−3	−2	−1	0	+1	+2	+3
Hair					•	•⋮•	
Brain		•	•		• •	•	• ••
Spleen		•			• •	• •	
Scales				• • • •			
Testes						•	
Ovary					•		
Thyroid					•		
Lungs				• • •	•	•	
Blood plasma						•	
Red blood cells				•	•		
Muscle				✕			
Heart				⋮ ⋮	• ••		
Fins				• • ••			
Liver			• •	⋮•• ⋮⋮ ••	⋮		
Whole			• •	⋮•• ⋮	•		•
Pancreas			•	⋮			
Chitin				•			
Kidneys				• ••	•		
Roe/eggs			• • •	•	•		
Feces			• •	• ⋮	•		
Thymus			•				
Fat	•	• •	•				
Hepatopancreas		•					

[a] Data are for a variety of terrestrial and marine animals (mice, men, gerbils, turtles, fish, shrimp, bivalves, and insects).

[b] Data are taken from DeNiro and Epstein (1978), Fry (1977, 1981), Fry *et al.* (1987), Gearing and Gearing (1991), Lyon and Baxter (1978), Rounick and Hicks (1985), Tieszen *et al.* (1983), and Yoneyama *et al.* (1983).

than a chemical, and some storage organs should change relatively slowly. For larger animals, most studies have used muscle tissue for analysis. Hair is another easily obtained tissue. It is distinctly heavier than most other tissues, making comparison with whole animals difficult, but it has the advantage of being able to show time trends. For smaller animals it is increasingly difficult to excise organs, and for animals less than around 1 cm, the whole animal is often used. There is very little difference between whole animal and muscle except in cases in which considerable quantities of gut contents or fat are present.

Specifically omitted from Table 3 are inorganic parts such as shell or bone. Inorganic carbon is much more positive than organic carbon (up to about 30 ‰). Any carbonate present in a sample must be removed before analysis, usually by acidification.

It is striking when looking at Fig. 1 to note how few analyses of individual organs can be found in the literature. The same can be said of biochemical differences, changes with sex and size, and most of the other parameters discussed in this chapter. For the present, researchers must trust the limited data available and take all reasonable precautions. More work on such variability, however, is needed to assure the best interpretation of $\delta^{13}C$ studies.

E. Seasonal Differences

Seasonal differences in ecosystems have been noted by several workers (Fry *et al.*, 1984; Gearing and Gearing, 1991). They may result from a combina-

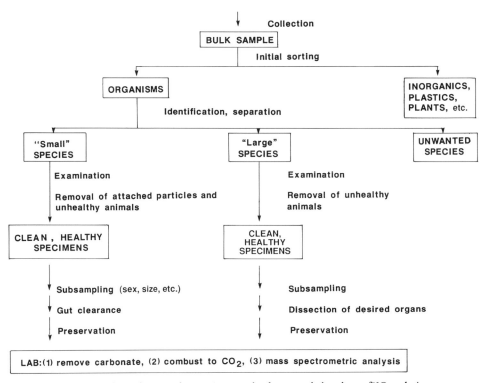

Fig. 1   Protocol for isolating and preparing samples for natural abundance $\delta^{13}C$ analysis.

tion of the parameters discussed above as well as from changes in the potential food physically present at different seasons (changing phytoplankton species, migrating animals, etc.). There is little evidence that temperature plays any *direct* role in causing isotopic seasonal differences. For studies of multiyear trends, extraneous variability can be minimized by sampling the same species at the same season. Otherwise, some effort should be made to examine animals throughout the year.

III. PROCEDURES

Principles and techniques will be given in this section for planning the collection of species as well as for isolating and purifying samples in the field. They are summarized in Fig. 1. Specific procedures will vary depending on the ecosystem being studied; marine ecosystems are used as examples herein. Laboratory procedures for combusting and analyzing the samples are given in Chapter 10, this volume.

A. Experimental Design

In order to most efficiently analyze an unknown ecosystem, a good general strategy is to progress from a broad survey to more specific questions. This applies to isotopic as well as other types of studies.

The first step should be to test the general, initial hypotheses by a trial survey of the area. This should include multiple analyses of primary producers and 10–20 $\delta^{13}C$ measurements on composite samples of the most important consumers. Questions to be asked include: (1) Are there isotopically distinct sources of organic carbon? (2) How different are these sources isotopically? (3) How do the isotope ratios of consumers compare with those of presumed sources, taking into account variation with trophic position? Are there any indications of unmeasured primary producers? (4) How does the variability between samples compare with the variability between sources? Compare these results with other information about the particular locale and with isotopic results (typical values and variability) from related areas. On the basis of this information, estimate the approximate cost (number of analyses needed for required degree of probability) to answer the questions posed and decide whether a full study is feasible. In some areas where the isotopic differences are small ($<2–3$ ‰) more samples are needed for statistical significance and the study may not be cost effective.

At this stage, specific hypotheses can be formulated. If a complete ecosystem study is appropriate, plan 2–3 sampling trips within a year. On each trip, collect 5–10 individuals of the important species and composite samples of other species. In general, it is better to collect 2 to 5 times more samples than you plan to analyze. Archived samples provide a backup for answering the questions that inevitably arise during the course of the experiment.

If the study is to be focused on the functioning of particular species, 10–20 individuals should be collected four or more times throughout their annual cycle or life cycle. Care should be taken to test possible sources of variability by collecting individuals of various sizes, particularly juveniles, nonbreeding adults, breeding adults of both sexes, and older animals.

If the study focuses on some particular food sources separated spatially, transects of stations may be appropriate. In these cases (for example tracing the extent of terrestrial or sewage carbon in surrounding ecosystems), it is important to measure the two end members several times during the study.

In all these studies, some measurements of inorganic carbon (air or DIC) should be made. Inorganic carbon is the foundation of the carbon isotope food web and changes due to burning fossil fuel or intense primary production can rapidly ($<$ one month) cause changes throughout the ecosystem.

B. Presampling

Before any sampling is done, information must be gathered and choices made in order to maximize the usefulness of the data in answering the specific questions addressed by the study. As stated earlier, no single kind of data should be used alone, particularly in ecological studies. Physical information should be gathered about the geography and topography of the general area, sources of water and water movements (currents, tides, etc.), climate (temperature, light levels, etc.), and history (anthropogenic inputs of organic matter, land use, introduction of different species, etc.). This will be used primarily to determine specific sampling sites. To this must be added biological and chemical information concerning the ecosystem. Data on hypothesized feeding relationships, plant succession, relative biomass and productivity of different species are all useful for choosing appropriate species for study. If certain species are chosen for intensive study, detailed knowledge of life history (seasonal biochemical changes due to food changes, spawning, overwintering forms, etc.) is necessary for choosing the particular times of sampling (see Section II).

Finally, a method of collection must be found for each species so as to obtain a group of individuals representative of the population in the ecosystem. The biological literature should be consulted for particular types of samplers for different types of ecosystems and populations. For examining relationships between species at a single site, the samples should be representative of the population as a whole in regard to such parameters as sex, size, health, age, and others (see Section II). For comparing two different areas, skewed groups can be tolerated as long as they are compared with a similar group in the other locations (all males of one species with all males of the same species, adolescents with adolescents, etc.).

C. Sampling and Separation

Sampling is simply a matter of applying the method chosen (see above) while noting the physical data available (date, time, immediate and long-term weather conditions, etc). Complexity arises in adapting to field conditions. Separation and cleaning of specimens is best done in the field immediately after collection while the animals are alive.

For the actual analysis, very little material (mg) is needed. However, enough individuals must be examined to assure that they represent an average of the whole population. Depending on the species density, it may be difficult to obtain enough specimens in what seems a reasonable length of time, but the subsequent analyses and their interpretation can only be as good as the initial set of samples.

Marine plankton are collected by filtering water through nets of varying sizes. However, the organisms tend to be irregular in shape, often with long spines, leading to separation by the longest dimension rather than true size. Towed nets of known mesh size also tend to become clogged, allowing the unquantitative collection of smaller organisms. Water samples can be collected at the same time for measurement of the $\delta^{13}C$ in dissolved inorganic carbon (250 ml water + 1 ml saturated mercuric chloride solution sealed tightly in a stoppered bottle and stored in a cool place). Isotope ratios of dissolved organic carbon are relatively invariate (1 – 3 ‰) in the open ocean but can vary over a larger range in estuaries (see Chapter 11, this volume).

After collection, the first step is removal of any inorganic or organic detritus present with the organisms. For larger animals this is relatively straightforward. For smaller aquatic and sedimentary animals, immediate, live microscopic investigation is best to eliminate, for example, bits of minerals, plants, and plastic.

Next the desired species must be identified and separated. It has been assumed in this chapter that a species is the basic group to be examined. This is not always the case. It is not always possible to collect enough identified animals at each site and sampling time. However, variability is increased by using larger groups. For example, in a study of the use of mangrove carbon in Malaysia, Rodelli *et al.* (1984) measured some crabs identified by species and others only by genus or higher group. Figure 2 shows the improvement in both accurate averages and in standard deviation obtained by analyzing the most specific category possible.

Separation becomes more difficult for smaller organisms, especially since it should be done with live animals and thus usually in the field. If species cannot be separated, they should be examined microscopically to determine their approximate composition. A subsample can be preserved for later identification, but some smaller organisms do not preserve well. Check for published procedures appropriate to the specific type of sample.

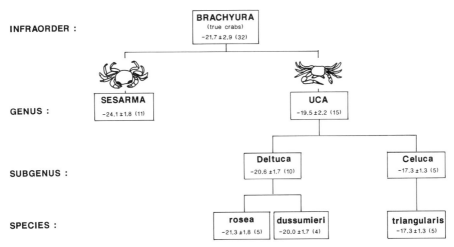

INFRAORDER :   **BRACHYURA** (true crabs) −21.7 ± 2.9 (32)

GENUS :   **SESARMA** −24.1 ± 1.8 (11)   **UCA** −19.5 ± 2.2 (15)

SUBGENUS :   **Deltuca** −20.6 ± 1.7 (10)   **Celuca** −17.3 ± 1.3 (5)

SPECIES :   **rosea** −21.3 ± 1.8 (5)   **dussumieri** −20.0 ± 1.7 (4)   **triangularis** −17.3 ± 1.3 (5)

Fig. 2   Change in $\delta^{13}C$ and its variability with taxonomic groups. Numbers give average ± one standard deviation (number of samples). Values are from Rodelli *et al.*, 1984.

For small sedimentary organisms, sieving can remove most mineral material, but there is no substitute for live sorting under a microscope. These organisms must be examined closely because many produce slime, which holds silt and clay particles around their bodies.

Plankton tows often contain a wide size range of organisms. Larger organisms such as small fish (up to several centimeters) and crustacea (shrimp, euphausiids, mysids) can be separated with a large mesh net or hand-picked with tweezers. All animals to be saved for analysis should be transferred to clean containers of filtered seawater. Do not hold carnivores or omnivores together, to prevent feeding. Small aquatic organisms such as zooplankton can often be separated from phytoplankton by sieving through various sizes of Nitex netting or by decanting (phytoplankton usually sink). Mineral particles can often be pipetted from the bottom of a beaker after swirling the water. These, like all particles, should be examined under the microscope before being thrown away because some gastropod larvae are often found in this fraction. Some zooplankton such as cladocerans and barnacle larvae float at the surface where they can be skimmed or pipetted off. There is no one perfect separation scheme because it depends on the animals initially present.

The individuals of desired species should then be examined to determine size, sex, approximate age, and general appearance. Different categories of the same species may be grouped together at this stage, for example, several females within certain size ranges. Whether analyzed separately or together, this basic information will be useful in case some numbers seem anomalous.

Of particular importance is the state of health or general appearance. Stressed or generally unhealthy individuals are almost certain to have a different biochemical composition (for example less fat) and this may change their $\delta^{13}C$ values (see Section II). Dead animals are also likely to differ in their biochemistry, for example, through decomposition, or leaching of more soluble compounds.

D. Purification

Like separation and identification, purification is best done on live animals as soon as possible after collection. Thus, some compromises are often necessary for field conditions. For large animals, this usually involves killing and dissection of the desired parts (usually muscle, see Section II) before storage.

Small animals that are to be analyzed whole (less than around 1 cm) must be examined microscopically and any adhering particles removed. This is particularly important for worms and other sedimentary organisms. They are then held live for gut clearance (the contents can change the isotope ratio, see Section II, Table 2). Aquatic animals are usually held in filtered seawater for several hours or overnight (change the water periodically to remove feces or let them fall through a mesh so as to separate from the animal). Small terrestrial animals should be held without food for the same period before killing.

Either at this stage or in the laboratory after proper storage (see next section), any inorganic carbon (isotopically more positive) should be removed with acid. This should not be a problem with pieces of muscle or with large animals. Small animals should always be tested for the presence of inorganic carbon by treating an aliquot with HCl. Animals such as copepods, mysids, and euphausiids may need to be powdered before acidification to remove all traces of inorganic carbon. The acid may dissolve some organic carbon from the organism. Although this has not been quantified, acidification is often carried out in the combustion vessel and the acid evaporated before combustion to avoid the loss of solubilized organic material.

E. Storage

Proper storage involves the conservation of the original isotopic abundance without changes due to added contaminants or losses during chemical or biological degradation. It is often most convenient to filter aquatic organisms before preservation to reduce the bulk. It should be noted that filtration may break up small cells, allowing some dissolved organic matter to be lost in the filtrate. Filters should be free of any organic carbon; glass fiber filters that have been cleaned by heating for over 4 hours at 500°C are usually used. Preservation methods include freezing and drying to reduce the risk of decomposition by bacteria. Formalin has been used as a preservative, but this treatment may allow some leaching out of more soluble organics and the leaching in of the formalin; it has been found to change the $\delta^{13}C$ value by around 1 to 2 ‰ (Mullin *et al.*, 1984). Although formalin-preserved samples have been used for analysis when better samples were not available, preservation in organic solvents is not a recommended method for isotopic samples. Freezing is the simplest and most rapid method if available in the field. In many out-of-the-way locations, drying is the only choice. Field samples can often be closed immediately in bags with silica gel to begin drying. A compromise must be made when drying between rapidity (fastest arrest of bacterial action) and loss of volatile organics (possibly changing the $\delta^{13}C$). Temperatures of 40 to 80°C have been used. Care must be taken to dry samples completely so as to avoid "smelly" situations later in the laboratory. All samples should be stored in containers free of oils or other available organic carbons.

F. Biochemical Separation

Separating individual compounds or classes of biochemicals for isotopic analysis makes it possible for natural isotope abundances to answer many new questions concerning chemical origins and biochemical pathways. Mass spectrometric technology has advanced to the point that individual compounds can be automatically separated by chromatography, combusted, and analyzed with precision better than 0.5 ‰.

For these measurements, the overriding consideration must be to avoid fractionating carbon isotopes during workup. Standard compounds should be put through the analytical procedure to quantify the isotopic changes resulting from the separation process.

Standard biochemical separation schemes are generally used. However, few isotopic tests of these procedures have been made. Because these techniques are not routine for $\delta^{13}C$ measurements, no detailed protocols are given here. Procedures can be found in various articles (for example, Degens *et al.*, 1968; DeNiro and Epstein, 1978; Des Marais *et al.*, 1980; Engel and Macko, 1986; Hayes *et al.*, 1987), but they should always be verified in each laboratory for isotopic discrimination.

G. Calculations

Natural abundances of carbon-13, used alone, can distinguish only two carbon sources. Since natural ecosystems are more complex than this, several different isotopes may be measured in order to distinguish multiple sources.

Calculations are based on simple mixing equations such as the following, which has been long used for carbon:

$$(X + Y)\,(\delta^{13}C_m) = (X)\,(\delta^{13}C_x) + (Y)\,(\delta^{13}C_y)$$

where

X = amount of carbon from source X,
Y = amount of carbon from source Y,
$\delta^{13}C_m$ = isotope ratio of the mixture,
$\delta^{13}C_x$ = isotope ratio of source X, and
$\delta^{13}C_y$ = isotope ratio of source Y.

Assuming only two sources of carbon, this can be rearranged to give:

$$F_x = \frac{\delta^{13}C_m - \delta^{13}C_y}{\delta^{13}C_x - \delta^{13}C_y}$$

where F_x is the fraction of carbon from source X.

The above equations were formulated for plants; they are usually used for animals, with sometimes a correction for trophic position. Accuracy can be improved by applying some modifications to take into account, for example, biochemical changes and seasonal variation. It is also best for $\delta^{13}C_x$ and $\delta^{13}C_y$ to use values for the same species of animal, collected at the same time, and known to have consumed only one source of food. This is not practical and often not possible, but reasonable efforts to improve the values of end members used will increase the accuracy of results. For example, McConnaughey and McRoy (1979) measured the amount of lipid in samples and corrected the isotope ratios for variations in this biochemical. Gearing *et al.* (1991) found that the addition of a second (pollutant) carbon source caused the isotope ratio of the original (natural) source to change due to changes in species and growth conditions; they made corrections for this effect as well as for trophic position (1% per trophic level). They also noted that pollutant carbon may enter and be carried through the food web in a manner different to that of the natural primary production.

REFERENCES

Anderson, R. K., Parker, P. L., and Lawrence, A. (1987). A $^{13}C/^{12}C$ tracer study of the utilization of presented feed by a commercially important shrimp *Penaeus vannamei* in a pond growout system. *J. World Aquaculture Soc.* 18, 148–155.

Degens, E. T., Behrendt, M., Gotthardt, B., and Reppmann, E. (1968). Metabolic fractionation of carbon isotopes in marine plankton—II. Data on samples collected off the coasts of Peru and Ecuador. *Deep-Sea Res.* 15, 11–20.

Des Marais, D. J., Mitchell, J. M., Meinschein, W. G., and Hayes, J. M. (1980). The carbon isotopic biogeochemistry of individual hydrocarbons in bat guano and the ecology of insectivorous bats in the region of Carlsbad, New Mexico. *Geochim. Cosmochim. Acta* 44, 2075–2086.

DeNiro, M. J., and Epstein, S. (1978). Influence of diet on the distribution of carbon isotopes in animals. *Geochim. Cosmochim. Acta* 42, 495–506.

DeNiro, M. J., and Epstein, S. (1981). Influence of diet on the distribution of nitrogen isotopes in animals. *Geochim. Cosmochim. Acta* **45**, 341–351.

Engel, M. H., and Macko, S. A. (1986). Stable isotope evaluation of the origins of amino acids in fossils. *Nature (London)* **323**, 531–533.

Fry, B. D. (1977). Stable carbon isotope ratios — a tool for tracing food chains. Masters Thesis, University of Texas, Austin.

Fry, B. D. (1981). Tracing shrimp migrations and diets using natural variations in stable isotopes. Ph.D. Dissertation, University of Texas, Austin.

Fry, B., and Arnold, C. (1982). Rapid $^{13}C/^{12}C$ turnover during growth of brown shrimp *(Penaeus aztecus)*. *Oecologia* **54**, 200–204.

Fry, B., and Parker, P. L. (1979). Animal diet in Texas seagrass meadows: $\delta^{13}C$ evidence for the importance of benthic plants. *Estuar. Coastal Mar. Sci.* **8**, 499–509.

Fry, B., Anderson, R. K., Entzeroth, L., Byrd, J. L., and Parker, P. L. (1984). ^{13}C enrichment and oceanic food web structure in the northwestern Gulf of Mexico. *Contrib. Mar. Sci.* **27**, 49–63.

Fry, B., Macko, S. A., and Zieman, J. C. (1987). Review of stable isotopic investigations of food webs in seagrass meadows. In "Subtropical-tropical Seagrasses in the Southeastern U.S." (M. J. Durako, R. C. Phillips, and R. R. Lewis, eds.), pp. 189–209. Florida Dept. Natural Res.

Gearing, J. N. (1988). The use of stable isotope ratios for tracing the nearshore-offshore exchange of organic matter. In "Lecture Notes on Coastal and Estuarine Studies, Vol. 22" (B.-O. Jansson, ed.), pp. 69–101. Springer-Verlag, Berlin.

Gearing, P. J., and Gearing, J. N. (1991). Isotopic variability of organic carbon in a phytoplankton-based, temperate estuary — II. Seasonal differences. (In preparation.)

Gearing, J. N., Gearing, P. J., Rudnick, D. T., Requejo, A. G., and Hutchins, M. J. (1984). Isotopic variability of organic carbon in a phytoplankton-based, temperate estuary. *Geochim. Cosmochim. Acta* **48**, 1089–1098.

Gearing, P. J., Gearing, J. N., Maughan, J. T., and Oviatt, C. A. (1990) Isotopic distribution of carbon from sewage sludge and eutrophication in the sediments and food web of estuarine ecosystems. *Environ. Sci. Technol.* (1991) **25**, 295–301.

Hayes, J. M., Takigiku, R., Ocampo, R., Callot, H. J., and Albrecht, P. (1987). Isotopic compositions and probable origins of organic molecules in the Eocene Messel shale. *Nature (London)* **329**, 48–51.

Ince, L. S., Mayer, L. M., Sherr, E. G., and Macko, S. A. (1982). Carbon inputs to bivalve mollusks: a comparison of two estuaries. *Can. J. Fish. Aquat. Sci.* **39**, 1348–1352.

Jones, R. J., Ludlow, M. M., Troughton, J. H., and Blunt, C. G. (1981). Changes in the natural carbon isotope ratios of the hair from steers fed diets of C4, C3 and C4 species in sequence. *Search* **12**, 85–87.

Kennicutt, M. C., Brooks, J. M., Bidigare, R. R., Fay, R. R., Wade, T. L., and McDonald, T. J. (1985). Vent-type taxa in a hydrocarbon seep region of the Louisiana slope. *Nature (London)* **317**, 351–353.

Kitting, C. L., Fry, B., and Morgan, M. D. (1984). Detection of inconspicuous epiphytic algae supporting food webs in seagrass meadows. *Oecologia* **62**, 145–149.

Leavitt, S. W., and Long, A. (1982). Evidence for $^{13}C/^{12}C$ fractionation between tree leaves and wood. *Nature (London)* **298**, 742–744.

Lyon, T. D. B., and Baxter, M. S. (1978). Stable carbon isotopes in human tissues. *Nature (London)* **273**, 750–751.

McConnaughey, T., and McRoy, C. P. (1979). Food web structure and the fractionation of carbon isotopes in the Bering Sea. *Mar. Biol.* **53**, 257–262.

Mullin, M. M., Rau, G. H., and Eppley, R. W. (1984). Stable nitrogen isotopes in zooplankton:

some geographic and temporal variations in the North Pacific. *Limnol. Oceanogr.* **29**, 1267–1273.

Newman, J. W., Parker, P. L., and Behrens, E. W. (1973). Stable carbon isotope ratios in Quaternary cores from the Gulf of Mexico. *Geochim. Cosmochim. Acta* **37**, 225–238.

Oviatt, C. A., Quinn, J. G., Maughan, J. T., Ellis, J. T., Sullivan, B. K., Gearing, J. N., Gearing, P. J., Hunt, C. D., Sampou, P. A., and Latimer, J. S. (1987). Fate and effects of sewage sludge in the coastal marine environment: a mesocosm experiment. *Mar. Ecol. Prog. Ser.* **41**, 187–203.

Peterson, B. J., and Fry, B. (1987). Stable isotopes in ecosystem studies. *Ann. Rev. Ecol. Syst. 1987* **18**, 293–320.

Peterson, B. J., and Howarth, R. W. (1987). Sulfur, carbon, and nitrogen isotopes used to trace organic matter flow in the salt-marsh estuaries of Sapelo Island, Georgia. *Limnol. Oceanogr.* **32**, 1195–1213.

Peterson, B., Fry, B., Deegan, L., Hershey, A., and Schell, D. (1991). Tracer analysis of new production in a Tundra River. (Submitted.) *Limnol. Oceanogr.*

Rodelli, M. R., Gearing, J. N., Gearing, P. J., Marshall, N., and Sasekumar, A. (1984). Stable isotope ratio as a tracer of mangrove carbon in Malaysian ecosystems. *Oecologia* **61**, 326–333.

Rounick, J. S., and Hicks, B. J. (1985). The stable carbon isotope ratios of fish and their invertebrate prey in four New Zealand rivers. *Freshwater Biol.* **15**, 207–214.

Rounick, J. S., and Winterbourn, M. J. (1986). Stable carbon isotopes and carbon flow in ecosystems. *Bioscience* **36**, 171–177.

Schoeninger, M. J., and DeNiro, M.J. (1984). Nitrogen and carbon isotopic composition of bone collagen from marine and terrestrial animals. *Geochim. Cosmochim. Acta* **48**, 625–639.

Spies, R. B., and Des Marais, D. J. (1983). Natural isotope study of trophic enrichment of marine benthic communities by petroleum seepage. *Mar. Biol.* **73**, 67–71.

Stephenson, R. L., Tan, F. C., and Mann, K. H. (1984). Stable carbon isotope variability in marine macrophytes and its implications for food web studies. *Mar. Biol.* **81**, 223–230.

Teeri, J. A., and Schoeller, D. A. (1979). $\delta^{13}C$ values of an herbivore and the ratio of C_3 to C_4 plant carbon in its diet. *Oecologia* **39**, 197–200.

Tieszen, L. L., Hein, D., Qvortrup, S. A., Troughton, J. H., and Imbamba, S. K. (1979). Use of $\delta^{13}C$ values to determine vegetation selectivity in east African herbivores. *Oecologia* **37**, 351–359.

Tieszen, L. L., Boutton, T. W., Tesdahl, K. G., and Slade, N. A. (1983). Fractionation and turnover of stable carbon isotopes in animal tissues: implications for $\delta^{13}C$ analysis of diet. *Oecologia* **57**, 32–37.

Tuross, N., Fogel, M. L., and Hare, P. E. (1988). Variability in the preservation of the isotopic composition of collagen from fossil bone. *Geochim. Cosmochim. Acta* **52**, 929–935.

van der Marwe, N. J. (1982). Carbon isotopes, photosynthesis, and archaeology. *Amer. Sci.* **70**, 596–606.

Yoneyama, T., Ohta, Y., and Ohtani, T. (1983). Variations of natural ^{13}C and ^{15}N abundances in the rat tissues and their correlation. *Radioisotopes* **32**, 330–332.

14

Tracer Studies with ¹³C-Enriched Substrates: Humans and Large Animals

Thomas W. Boutton

Department of Rangeland Ecology and Management
Texas Agricultural Experiment Station
Texas A&M University
College Station, TX 77843

I. INTRODUCTION

A. Use of Stable Carbon Isotopes in Metabolic Research

The use of stable carbon isotopes in metabolic research on humans and large animals is expanding rapidly due to the increasing variety of labeled compounds, greater availability of analytical facilities, and absence of health risk from radiation. Studies on humans, especially those on children and women of child-bearing age, now rely almost exclusively on ¹³C- rather than ¹⁴C-labeled substrates. While substrate and analytical costs remain slightly higher for studies with ¹³C compared with ¹⁴C (Klein *et al.*, 1981), the risks associated with radioisotope usage often preclude their use in many segments of the human population.

To date, few large animal studies have used ¹³C tracers because of the cost of the massive doses required by large body size. However, the value of the large animals that must be destroyed at the end of radioactive tracer studies, the extensive federal regulations surrounding radioisotope usage, and pressure from animal rights groups may increase the use of stable carbon isotopes in studies on large animals. The realization that significant physiological information on large animals can be derived from simple and inexpensive manipulation of the natural ¹³C content of the animal's diet (e.g., Boutton *et al.*, 1988) may also increase stable isotope usage in animal studies.

In vivo studies of human and animal metabolism using ¹³C-labeled substrates can provide several types of information including: (1) presence and activity of specific enzymes; (2) absorption, malabsorption, and oxidization of dietary nutrients; (3) synthesis and degradation (turnover rates) of biochemicals; and (4) presence or absence of disease or metabolic disorders. These applications have been reviewed extensively (Bier and Matthews, 1982; Halliday and Rennie, 1982; Klein and Klein, 1985).

This chapter will focus on methodology required for *in vivo* studies aimed at quantifying the fate of dietary nutrients. This choice is dictated largely by the author's own experience, and by the current intense interest in both human and animal nutrition. Following an outline of the general methodology involved in these studies, a general protocol for studying utilization of ^{13}C starch by young infants will be described. It is assumed that the reader is familiar with the instrumentation and isotopic notation described in Chapter 10, this volume.

B.   Background Levels of ^{13}C

Studies with ^{13}C differ from those with ^{14}C in that there is a significant ($\sim 1.1\%$ of all carbon) natural background of ^{13}C. Thus, dietary nutrients, animal tissues, and animal excreta all contain ^{13}C. This natural background of ^{13}C in diet, tissues, and excreta is somewhat variable (1.08–1.10 atom %), and depends on: (1) the relative proportions of C_3 and C_4 plant foods (see Chapter 13, this volume) in the diet; (2) the amount of seafood in the diet; (3) the metabolic fuel mix at the time of sampling (lipids contain less ^{13}C than carbohydrate and protein); and (4) the physiological state of the subject (e.g., fasted or fed, rested or stressed) at the time of sampling (Wolfe *et al.*, 1984; Schoeller *et al.*, 1984). Individual food items cover a range of $\delta^{13}C_{PDB}$ values from approximately -10 ‰ (1.1002 atom %) to -30 ‰ (1.0783 atom %) (Nakamura *et al.*, 1982). Background levels of ^{13}C must be established in all sample types in order to resolve natural ^{13}C from excess ^{13}C derived from metabolism of the labeled substrate.

Variation in the ^{13}C background in respiratory CO_2 during the course of a test can be minimized by requiring subjects to fast during and for 12 hours before a test, and to remain at rest (preferably lying down) during the test. Under these conditions, background levels of ^{13}C in breath remain relatively constant (1 SD = 0.7 ‰) for the duration of the test, and can be determined reliably by a few predose samples (Schoeller *et al.*, 1977, 1980). However, some protocols require that the subject be fed and exercised during the test, resulting in substantial (up to 5 ‰) excursions of background ^{13}C in breath (Schoeller *et al.*, 1984; Wolfe *et al.*, 1984). In such cases, the protocol should be repeated on an alternate day with an identical feeding and exercise regime, but without the ^{13}C-labeled test substrate (Jones *et al.*, 1985). Background levels of ^{13}C obtained under these circumstances can then be used to correct ^{13}C levels obtained during an actual ^{13}C test.

The detection limits for excess ^{13}C in respiratory CO_2, stool, and urine have been considered with respect to background variability of ^{13}C in those excreta. In general, the minimum excretion rates required to detect a significant (2 × background SD) increase in ^{13}C above background for different sample types are: (1) 800 nM excess ^{13}C/kg body wt/24 hr for stool (Schoeller *et al.*, 1981); (2) 140 nM excess $^{13}CO_2$/kg body wt/hr for respiratory CO_2 (Schoeller *et al.*, 1977); and (3) 500 nM excess ^{13}C/24 hr for urine (von Unruh *et al.*, 1974).

C. Planning and Executing a Protocol Using ^{13}C-Labeled Substrates

Planning and executing an *in vivo* metabolic study using a ^{13}C-labeled substrate involves multiple considerations (Table 1). First, a question must be formulated and a sample population (e.g., lactating range cows, premature infants) defined. Then, decisions must be made regarding which labeled substrate will best answer the question at hand, how that substrate should be labeled (at specific molecular positions or uniformly), to what extent it should be enriched with ^{13}C, the size of the dose required to produce an

Table 1

Outline of Major Steps Required for Successful Completion of *in Vivo* Metabolic Studies with Stable Isotopes

I. Protocol Planning
    A. Formulation of question
    B. Definition of sample population
    C. Choice of appropriate labeled substrate
    D. Size of dose
    E. Route of administration of dose
    F. Mode of substrate delivery
    G. Types of samples to collect
    H. Timing and duration of sample collection
    I. Approval of human or animal use committees
II. Execution
    A. Determination of baseline ^{13}C levels
    B. Administration of labeled substrate
    C. Collection of samples
    D. Preservation of samples
III. Sample Analysis and Data Interpretation
    A. Sample preparation and isotopic analysis
    B. Analysis and modeling of data
    C. Interpretation

isotopic signal in the sample, the route of administration (oral or intravenous), the mode of administration (bolus or constant infusion), the kinds of samples to be collected, and the timing and duration of sample collection. Finally, a study plan must be approved by human or animal use committees, and informed consent must be obtained if human subjects are participating.

The second stage involves execution of the protocol (Table 1). Prior to administration of the ^{13}C-labeled substrate, predose or background levels of ^{13}C must be established for all sample types in order to be able to distinguish natural ^{13}C from excess ^{13}C originating from the labeled substrate. Then, the labeled substrate can be administered, and postdose samples collected. Samples must be preserved in a manner that prevents change in isotopic content during storage. For example, if stool samples are not frozen immediately, bacteria present in the stool may continue to metabolize malabsorbed substrate, resulting in possible differential loss of ^{13}C or ^{12}C from the sample and an incorrect estimate of substrate malabsorption.

In the third stage (Table 1), samples are prepared and analyzed for carbon isotope composition. Following calculations of substrate recovery in the different sample types, it may be possible to construct a model describing substrate kinetics. Results of the modeling procedure may be useful in directing additional research effort.

II. MATERIALS REQUIRED

A. Instrumentation for Measurement of Carbon Isotope Ratios

By far, the most often used technique for precise quantitation of ^{13}C tracers administered *in vivo* has been mass spectrometry. However, a choice must be made as to which type of mass spectrometer is best for a given application. The dual-inlet, triple-collector gas isotope ratio mass spectrometer (GIRMS) has been described in detail in Chapter 10, this volume. This instrument is best suited for applications where isotopic differences between samples are small and where sample size is not a problem; a minimum of 10 μl of CO_2 at STP (or 5 μg carbon) is required for a carbon isotope ratio measurement. Samples such as respiratory CO_2, stool, urine, milk, tissue, and whole blood will generally be large enough to exceed this requirement.

However, if the isotopic composition of a specific biochemical must be measured, such as blood glucose or lysine in milk, then meeting the sample size requirement for GIRMS becomes difficult or impossible. Instrumental requirements for this application are best met by gas chromatography/mass spectrometry (GC/MS), which can readily determine the isotopic composition of very small quantities (nmol or pmol) of specific biochemicals. For example, 100 μl of blood plasma is sufficient to permit isotopic measurements of specific free amino acids by GC/MS.

GC/MS and isotope ratio mass spectrometry also differ in the precision of the measured isotope ratios. Isotope ratio mass spectrometry can routinely provide precision (machine error plus sample preparation error) of 0.1 ‰ while that for GC/MS may range from 10 to 100 ‰. Thus, GIRMS is ideally suited for applications in which samples are large and isotopic differences between samples are small, while GC/MS is appropriate when sample sizes are small and isotopic differences between samples are large. GC/MS will not be considered further in this chapter, but the interested reader should consult Matthews and Bier (1983) and Hachey *et al.* (1987).

An infrared heterodyne ratiometer based on differential absorption of infrared wavelengths by $^{13}CO_2$ and $^{12}CO_2$ has been described and shown to be capable of analyzing the $^{13}C/^{12}C$ content of respiratory CO_2 with a precision of approximately 0.4 ‰ (Irving *et al.*, 1986). Because background or baseline levels of ^{13}C in respired CO_2 from rested, fasted individuals will vary by about 0.7 ‰ (1 SD) over a 6-hr period, the precision of this infrared-based instrument is acceptable for metabolic studies involving respiratory CO_2. Advantages of infrared heterodyne ratiometers include: (1) the potential for continuous on-line monitoring of $^{13}C/^{12}C$ in respiratory CO_2; (2) absence of sample preparation requirements for analysis; (3) absence of a high vacuum system; (4) minimal operator expertise and training; and (5) significantly lower cost relative to a mass spectrometer. However, instruments of this design are not yet available commercially.

B. ^{13}C-Labeled Test Substrates and Standards

Substrates highly enriched with ^{13}C (2–99 atom %) are available commercially from several sources. In general, the cost increases with the complexity of the molecule and the degree of isotopic enrichment. The metabolic pathway followed by the substrate and the extent to which the substrate or its by-products are diluted in the sample will determine where in the molecule the ^{13}C label must be located, and the extent to which it must be enriched. The label can be located at a specific position within the substrate molecule (e.g., C_1 in glucose), or all carbon atoms within the substrate can be uniformly labeled. While simpler ^{13}C substrates such as sugars, amino acids, and fatty acids are readily available, more complex molecules such as proteins and carbohydrates generally are not. However, these complex biochemicals can be labeled by providing photosynthetic organisms with $^{13}CO_2$ during photosynthesis, and then isolating the compound of interest (Kollman *et al.*, 1973; Boutton *et al.*, 1987). Details of ^{13}C labeling of plants have been presented elsewhere (Svejcar *et al.*, 1990).

In addition to substrates highly enriched with ^{13}C, it is also possible to use naturally enriched test substrates in metabolic studies. For example, plants with the C_4 pathway of photosynthesis (such as corn, sugar cane, sorghum, and other tropical grasses) are naturally enriched in ^{13}C compared with C_3 plants. Because most of the human diet consists of food items derived from C_3 plants, the metabolic utilization of a dose of cane sugar or corn starch can be assessed quantitatively by monitoring the increase of ^{13}C in respiratory CO_2 (Lefebvre, 1985; Shulman *et al.*, 1986). Similarly, significant physiological information can be obtained from large animals by simple manipulation of the natural ^{13}C content of the diet (Tieszen *et al.*, 1983; Boutton *et al.*, 1988; Wilson *et al.*, 1988; Metges *et al.*, 1990).

The dose required to conduct a metabolic study with a ^{13}C-labeled substrate will depend upon body size, the ^{13}C enrichment of the substrate, the rate at which the substrate is metabolized, and the nature of the samples to be collected. For substrates enriched with ^{13}C to approximately 90 atom % at one or more positions, doses will typically range from 2 to 10 mg of substrate per kg of body weight. For naturally enriched substrates (e.g., corn glucose) with $\delta^{13}C$ values of approximately -12 ‰ versus PDB, doses of 1 g/kg of body weight are usually adequate.

Standards for carbon isotope analysis by isotope ratio mass spectrometry have been described in Chapter 10, this volume. All of the standards described in that chapter have $\delta^{13}C$ values at natural abundance levels. However, four new ^{13}C-enriched reference materials are available from the International Atomic Energy Agency in Vienna (Anonymous, 1987). Two of the reference materials are uniformly labeled glucose (~ 100 ‰ and 600 ‰ vs PDB), and two are $NaHCO_3$ solutions (~ 100 ‰ and 500 ‰ vs PDB). The $\delta^{13}C$ values of these new reference materials have only been estimated, and a program is underway at IAEA to establish precise values by consensus of the laboratories participating in the calibration.

C. Sample Collection Equipment

Metabolic studies aimed at assessing utilization of ^{13}C-labeled nutrients will often involve samples of respiratory CO_2, stool, urine, milk, blood, or tissue. The most important consideration for all sample types is the preservation of isotopic integrity during collection, storage, and sample preparation.

1. Collection of Respiratory Carbon Dioxide

Oxidation of the ^{13}C-labeled substrate for energy production by humans or animals is assessed by collection of respiratory CO_2 for isotopic analysis at specified time intervals for several hours postdose. The increase in $^{13}CO_2$ in the breath together with total CO_2 production are then used to calculate the proportion of the substrate oxidized for energy production. Perhaps the simplest method for collecting respiratory CO_2 is with a face mask system (Fig. 1). A standard resuscitation face mask covering both nose and mouth

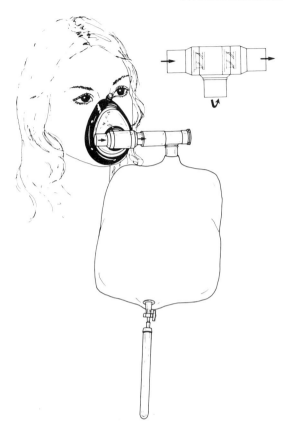

Fig. 1   Face mask system for collection of respiratory CO_2 for carbon isotope analysis. Close up shows details of two-way nonrebreathing valve.

(e.g., Vital Signs, Inc., Totowa, New Jersey) is fitted with a two-way nonrebreathing valve (Hans Rudolph, Inc., Kansas City, Missouri). These valves are available in a variety of sizes capable of accommodating humans from birth through adulthood, and animals ranging in size from rats to horses. For humans, a gas-tight 3-l collection bag is attached to the outlet portion of the nonrebreathing valve to capture respired air. The collection bag should be equipped with a sample port, to which a one-way stopcock and a syringe needle are attached (Fig. 1). Similar face mask systems for collecting respired air have been constructed for use with large animals (Chevalier *et al.,* 1984). When the collection bag is full, the stopcock can be opened, and the syringe needle pushed through the rubber septum on a 20-ml Vacutainer® (Becton Dickinson Co., Rutherford, New Jersey) or Venoject® tube (Terumo Medical Corp., Elkton, Maryland). Because the Vacutainer is under vacuum, a 20-ml aliquot of the breath sample is pulled into the tube. To avoid isotopic memory effects, the breath collection bag should be thoroughly emptied between samples by detaching the bag from the face mask assembly (Fig. 1) and flattening the bag completely. The large volume of respiratory gas (3 l) collected in the bag during each sample ensures that memory is not a problem if this simple precaution is taken.

Samples collected and stored in Vacutainers retain their isotopic composition for at least 3 months (Schoeller and Klein, 1978). The Vacutainers should be non-silicone coated and nonsterile (available by special order from Becton Dickinson or Terumo) to avoid possible contamination from residual gases (Ajami and Watkins, 1983). The problem of residual gas contamination has been shown to be more serious in Vacutainers than in Venoject tubes (Milne and McGaw, 1987).

In some cases, the subject may be confined to a direct or indirect calorimeter for continuous measurement of respiratory quotient (RQ) and energy expenditure (Fig. 2). In this situation, the concentration of CO_2 in the effluent gas stream from the calorimeter is too dilute for collection in Vacutainers. Respiratory CO_2 must be collected by passing a portion of the effluent gas stream from the calorimeter through a $1N$ solution of carbonate-free NaOH, thereby concentrating and trapping the CO_2 as $NaHCO_3$. However, if CO_2 trapping is not 100% efficient, isotopic fractionation results. To ensure complete CO_2 trapping, a gas-washing device similar to that shown in Fig. 2 is recommended. This device forces gas bubbles to spiral slowly up through approximately 15 ml of $1N$ NaOH solution, prolonging contact between the gas bubbles and the NaOH. The NaOH used in the trap must be carbonate free (e.g., J.T. Baker Chemicals, Phillipsburg, New Jersey) to avoid large blanks, and the stock solution should be kept in a tightly closed container to prevent absorption of atmospheric CO_2.

Another important consideration with this method is the presence of atmospheric CO_2 in the calorimeter gas stream. This problem can be eliminated by scrubbing CO_2 from the air with ascarite before it enters the chamber, or by measuring the isotopic composition and flow rate of the atmospheric CO_2 entering the calorimeter and using mass balance calculations to correct for the isotopic contribution from atmosphere.

Carbon dioxide for isotopic analysis can be generated from the $NaHCO_3$ by reaction of an aliquot of the trapping solution with 85% phosphoric acid under vacuum, as described in Chapter 10, this volume. The amount of trapping solution needed to produce an adequate volume of CO_2 following acidification will depend on the CO_2 concentration of the calorimeter effluent gas, the flow rate of the gas passing through the NaOH trap, and the length of time over which CO_2 trapping occurred. For example, if a subsample of the air stream leaving the calorimeter (Fig. 2) contains 0.25% CO_2 and is passed through the gas washing device at 300 ml/min, then a 5-min

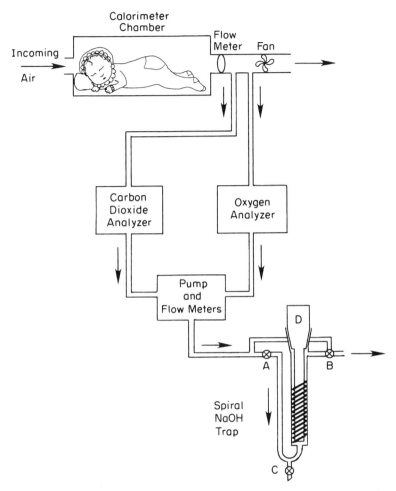

Fig. 2 An indirect calorimetry system coupled with a gas-washing device for trapping respiratory CO_2 in $1N$ NaOH solution. A subsample of the calorimeter gas stream enters the NaOH trap on the left side and is forced down into the NaOH solution, where bubbles are formed which must spiral around the solid center piece before exiting trap on right. Stopcocks A and B can be adjusted so gas stream bypasses trap while NaOH is drained through stopcock C at the end of a sample collection. Fresh NaOH can be added by removing center piece with spirals (D).

collection period will trap approximately 0.15 mmol of carbon in the 15 ml of $1N$ NaOH. Therefore, to produce a 1-ml (0.04 mmol) sample of CO_2 for mass spectrometric analysis, approximately 4.5 ml of the NaOH trapping solution should be acidified. While ¹³C substrate studies on subjects undergoing indirect calorimetry are logistically more difficult, the information derived from RQ and energy expenditure measurements is often complementary to that derived from ^{13}C substrate oxidation data (e.g., Sauer *et al.,* 1986). Additional techniques for collection of respiratory CO_2 can be found in Wolfe (1984).

2. Other Samples

When malabsorption of a ^{13}C-labeled substrate is being assessed, all stool samples must be collected quantitatively so that both stool mass and isotopic composition can be measured accurately over the time period of interest. Any method for accomplishing this is acceptable, and no special equipment is needed. Stool should be frozen immediately after collection to prevent loss of volatile fatty acids and to prevent bacterial metabolism of the malabsorbed substrate. Blood, milk, and urine do not require special equipment, but samples should be stored frozen to prevent decomposition.

D. Equipment for Measuring Total Respiratory CO_2 Production

In order to calculate the recovery of ^{13}C in breath, both the isotopic composition and rate of production of CO_2 must be known. The simplest method for obtaining total CO_2 production (VCO_2) for humans is by estimation from body weight or body surface area. For weight, the relationship usually employed is 9 mmol CO_2/kg body wt/hr (Winchell *et al.,* 1970), while that for body surface area is 5 mmol/m²/min (Shreeve *et al.,* 1970). Body surface area (BSA) can be calculated from height and weight measurements (Haycock *et al.,* 1978) as

$$BSA(m^2) = wt(kg)^{0.5378} \times ht(cm)^{0.3964} \times 0.024265 \qquad (1)$$

Although these relationships based on weight or body surface area provide useful approximations for fasted individuals at rest, they cannot account for effects of feeding, exercise, or physiological state on the rate of CO_2 production.

Total CO_2 production can be measured directly using the face mask system shown in Fig. 3, which is only a slight modification of that used to collect breath for isotopic analysis. A standard resuscitation face mask is fitted with a two-way nonrebreathing valve. A mechanical Wright respirometer (Fraser-Harlake, Orcano Park, New Jersey) is attached to the inlet side of the two-way valve with a short length of flexible aerosol tubing (Vacuumed, Ventura, California) to measure respiratory flow. A gas-tight bag (Vacuumed, Ventura, California) with a volume of 3 to 20 l (depending on length of collection period) is attached to the outlet side of the valve, and captures expired air. Samples of expired air can be withdrawn from the bag

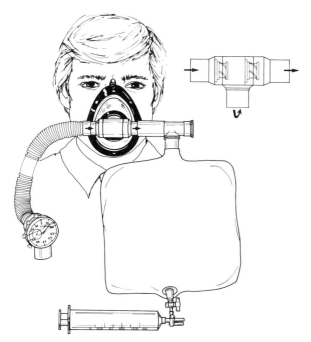

Fig. 3 Face mask system for measurement of total CO_2 production. A respirometer is connected to the two-way nonrebreathing valve with a short length of flexible tubing. Aliquots of respired air are withdrawn from the gas-tight bag with 50-ml syringes, and measured for CO_2 concentration by gas chromatography or infrared gas analysis.

through a sample port equipped with a stopcock. Aliquots of expired air can be drawn from the bag with 50-ml syringes, and analyzed for CO_2 concentration by gas chromatography or infrared gas analysis. Gas samples should not be stored in syringes for more than 6 hr. Total CO_2 production is measured by recording respiratory air flow with the respirometer over a defined length of time (2–5 min is adequate), and by measuring the CO_2 concentration in the expired air collected in the bag during that same time interval. The product of the respiratory volume per unit of time and the CO_2 concentration of that volume yields the CO_2 production rate. Measurements of CO_2 production rate using a face mask system have been shown to compare favorably with other more sophisticated and cumbersome methods (Segal, 1987). Similar systems have been described for large animals (McKirnan *et al.*, 1986). Total CO_2 production should be assessed periodically (hourly) for the duration of a study.

Respiratory CO_2 production can be measured continuously by indirect calorimetry using a system similar to that shown in Fig. 2. However, indirect calorimetry systems are expensive and require considerable expertise to use properly. A detailed description of indirect calorimetry is given in Flatt (1969).

E. Sample Preparation Equipment

1. Combustion of Organic Samples

Because isotope ratio mass spectrometers measure carbon isotope ratios on CO_2 gas only, all organic samples (stool, blood, urine, milk, tissue) must be combusted to CO_2 prior to analysis. Samples must be thoroughly homogenized, dried, and ground to a powder prior to combustion. Although some mass spectrometers with low-volume inlet systems can accommodate samples as small as 5 μg of carbon, a minimum of 1–2 mg of carbon is generally required for a single combustion. Details on a sealed-tube combustion procedure for organic samples are given in Chapter 10, this volume. In order to quantify the amount of label present in a sample, the % carbon of that sample must be known. The % carbon can be determined by measuring the weight of the aliquot combusted and calculating mg carbon in that aliquot from the volume of CO_2 produced by combustion (Chapter 10, this volume). If % carbon must be determined, the combustion technique must assure quantitative conversion of all sample carbon to CO_2 (Boutton *et al.*, 1983).

2. Purification of Respiratory CO_2

Carbon dioxide must be isolated from other respiratory gases by cryogenic distillation prior to mass spectrometric analysis. This can be accomplished using the vacuum system shown in Fig. 4. If the respiratory gases are

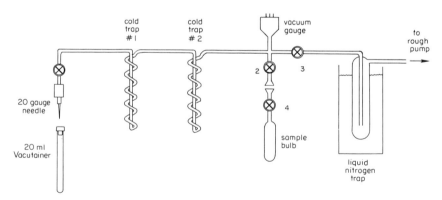

Fig. 4 Vacuum system for isolation and purification of respiratory CO_2 from breath samples collected in Vacutainers.® Cold traps 1 and 2 are approximately 20 cm in height, 6 cm in width, and constructed from 3/8-in o.d. glass tubing. Circles with Xs represent high-vacuum stopcocks.

collected and stored in Vacutainers, the vacuum system must be equipped with a needle (approximately 20 gauge) for puncturing the septum of the Vacutainer to admit the respiratory gases into the system. A 1-in stainless steel syringe needle with a stainless steel hub can be welded onto a 5-cm length of 1/4-in o.d. stainless steel tubing, and attached to the vacuum system with a Cajon ultratorr fitting (Cajon Company, Macedonia, Ohio). A noncoring syringe needle should be used to prevent the needle from clogging with rubber stopper.

First, a sample bulb is attached to the vacuum system, and valves 2, 3, and 4 are all opened to evacuate the system to $<10^{-2}$ torr. This can be accomplished with a mechanical vacuum pump attached to a liquid nitrogen trap to prevent back-streaming of oil vapor (Fig. 4). When a suitable vacuum has been achieved, a Vacutainer should be pushed only far enough onto the needle to completely cover the opening of the needle. If the end of the needle protrudes into the Vacutainer, respiratory gases will be pumped away prematurely. Then valve 1 can be opened and the dead space between valve 1 and the Vacutainer evacuated.

When a vacuum of $<10^{-2}$ torr has been achieved, liquid nitrogen Dewars are placed around cold traps 1 and 2, and valves 2 and 3 are closed. The Vacutainer is pushed upwards until the needle fully penetrates the septum, and the CO_2 and water are allowed to freeze into the traps for 4 min. Then, valve 3 is opened slightly to ensure that all respiratory gases are pulled out of the Vacutainer and through the traps. All CO_2 and water is frozen in cold trap 1, but cold trap 2 ensures that no CO_2 is pumped away. Nitrogen, oxygen, and other respiratory gases not frozen at liquid nitrogen temperature ($-196°C$) are pumped away. When a vacuum of $<10^{-2}$ torr has been reached, valves 1 and 3 are closed, 2 and 4 opened, the liquid nitrogen Dewars on cold traps 1 and 2 replaced with ethanol-dry ice Dewars ($\sim -86°C$), and the sample bulb immersed in liquid nitrogen. This allows the frozen CO_2 in cold traps 1 and 2 to sublime and diffuse into the sample bulb. The transfer of CO_2 can be monitored with the vacuum gauge, and can be considered complete when a vacuum of $<10^{-2}$ torr has been restored. At least 3 min should be allowed for the transfer. At this time, valves 4 and 2 are closed and the sample bulb is detached from the vacuum system and taken to the mass spectrometer for isotopic analysis. Since respired air is approximately 5% CO_2, a 20-ml Vacutainer sample will generally yield 1 ml of CO_2 at STP.

In preparation for the next sample, valve 3 is opened and the dry-ice Dewars removed from cold traps 1 and 2. A heat gun (400°C) should be used to heat cold traps 1 and 2 to drive out water vapor. Approximately 15 min is required to completely process one sample. Other procedures for purifying respiratory CO_2 are outlined by Schoeller and Klein (1978) and Chevalier *et al.* (1984).

A sample preparation system for Vacutainer samples similar to that described above has been automated and interfaced with a gas isotope ratio mass spectrometer (Schoeller and Klein, 1979), allowing unattended purification and analysis of approximately 75 breath samples per day. Instruments of similar design are now available commercially.

III. PROCEDURES

To illustrate the application of the techniques outlined above, a protocol dealing with utilization of ^{13}C-labeled starch by young children will be summarized with emphasis on procedures and calculations.

A. Utilization of ^{13}C-Labeled Starch by Infants and Young Children

1. Background

The capacity of human infants to digest complex carbohydrates is generally considered to be limited because pancreatic amylase, the most important enzyme for complex starch degradation, is not present for the first 6 months of life. However, mucosal amylase, salivary amylase, breast milk amylase, and bacterial fermentation in the colon may all provide alternate mechanisms by which complex starches can be utilized by infants prior to the development of pancreatic amylase activity. As a probe for studying starch utilization in this population, a ^{13}C-labeled starch was produced. A protocol for measuring oxidation and malabsorption of the labeled starch is described below.

2. ^{13}C-Labeled Substrate and Calculation of Dose

Because ^{13}C-labeled starch was not available commercially, rice plants were enclosed in plexiglass chambers and exposed to ^{13}CO$_2$ during the grain-filling portion of their life cycle (Boutton *et al.*, 1987). The plants were grown to maturity and milled to produce white rice. An infant cereal identical to a commercial product was produced from the ^{13}C-labeled rice by Gerber Products Company (Fremont, Michigan). Although chemical fractionation revealed that some of the labeled carbon in the cereal was present as rice protein and lipid, approximately 95% of the total carbon and 95% of the excess ^{13}C in the cereal was in the starch fraction. Furthermore, the amylose and amylopectin fractions of the starch, which differ slightly in digestibility, had identical ^{13}C/^{12}C ratios (Boutton *et al.*, 1987). Since protein and fat are not substrates for energy production when starch is available, the contribution from oxidation of protein and lipid should be negligible.

A general formula for calculating the mmol of ^{13}C in a dose is given in

Table 2
Formulas Needed for Calculation of ^{13}C Recovery in Respiratory CO_2

---

a. $\delta^{13}C_{PDB} = \left[\dfrac{R_{sample}}{R_{PDB}} - 1 \right] \times 1000$

b. $R_{sample} = \left[\dfrac{\delta^{13}C_{PDB}}{1000} + 1 \right] \times R_{PDB}{}^{a}$

c. $F = \dfrac{R_{sample}}{R_{sample} + 1}$

d. atom % excess (APE) $= (F_{postdose} - F_{baseline}) \times 100$

e. mmol excess ^{13}C in dose $= \dfrac{\text{amount given (mg)}}{\text{molecular weight}^{b}} \times \dfrac{APE}{100} \times \dfrac{\text{number of}}{\text{labeled atoms}}$

f. $\dfrac{\mu mol \text{ excess } ^{13}C}{\text{minute}} = \dfrac{APE \text{ breath}}{100} \times \dfrac{\mu mol \text{ total } CO_2}{\text{minute}}$

g. % dose/minute (PCD) $= \dfrac{\mu mol \text{ excess } ^{13}C/\text{min}}{\mu mol \text{ excess } ^{13}C \text{ in dose}} \times 100$

h. $\dfrac{\text{cumulative % dose}}{\text{recovered (CUMPCD)}} = \left[\dfrac{(PCD_{t} + PCD_{t-1}) \times \Delta t(\text{min})}{2} \right] + CUMPCD_{t-1}$

---

^{a}The actual value for R_{PDB} is 0.0112372.
^{b}Must be based on the molecular weight of the ^{13}C-enriched material, not the molecular weight at natural abundance levels of ^{13}C.

Table 2 (formula e). This formula can be used for simple substrates for which the molecular weight, isotopic enrichment, and number of labeled carbon atoms are known. However, because starch is complex and consists of many types of starches differing in chain length and molecular weight, the formula cannot be applied in this case. Nonetheless, by knowing the % carbon and isotopic composition of the cereal, the amount of ^{13}C in a given dose of cereal can be calculated. This cereal had a $\delta^{13}C_{PDB}$ of $+1098$ ‰ (or 2.3033 atom % ^{13}C) and was 40.7% carbon. The average atomic weight of carbon in the cereal was $(12 \times 0.976963) + (13 \times 0.023033) = 12.023$. Thus, 1 g of ^{13}C-labeled cereal would contain 0.407 g C/12.023 $=$ 33.852 mmol of total carbon. Some of the ^{13}C in the cereal is background or natural abundance ^{13}C. The $\delta^{13}C_{PDB}$ value for unlabeled rice cereal was found to be -24.2 ‰, or 1.0846 atom % ^{13}C. To determine the mmol of excess ^{13}C in the labeled cereal, the mmol of total carbon in the cereal is multiplied by the difference in fractional abundances of the labeled and unlabeled cereals. Therefore, accounting for the presence of background ^{13}C, a 1-g quantity of ^{13}C-labeled rice cereal would have an effective dose of 33.852 $\times$ (.023033 − .010846) $= 0.4126$ mmol of excess ^{13}C.

3. Subjects and Experimental Design

Subject 1 was a 3-month-old female with a weight of 6.1 kg and height of 61.5 cm. Body surface area was calculated (Haycock *et al.*, 1978) to be 0.3296 m². Subject 2 was a 3-year-old male included as a control subject with a weight of 12.3 kg, height of 94.0 cm, and a calculated (Haycock *et al.*, 1978) body surface area of 0.5666 m². Subject 1 had been formula fed since birth, and on the basis of age was presumed to have limited ability to digest starch. Subject 2 was fully weaned, had a normal human diet, and was presumed to have a mature digestive system.

Baseline stool samples were collected from both subjects for 24 hr prior to the test, and frozen immediately upon collection. Both subjects were fasted overnight (12 hr) to minimize baseline variation in respiratory $^{13}CO_2$ prior to receiving the labeled substrate, and were confined to bed for the duration of the study to minimize variation in total respiratory CO_2 production. Baseline breath samples were collected from both subjects using the face-mask system (Fig. 1) at 30, 15, and 2 min prior to receiving the labeled cereal. Subject 1 received 4.0563 g of labeled cereal (1.6736 mmol excess ^{13}C) dissolved in 120 ml of infant formula. To rinse any residual substrate, the bottle was refilled with 90 ml of infant formula, shaken thoroughly, and the contents consumed by subject 1. Subject 2 received 5.7745 g of labeled cereal (2.3826 mmol excess ^{13}C) dissolved in 180 ml of milk. As before, the cup was rinsed with an additional 180 ml of milk to ensure that all labeled cereal was consumed. Postdose breath samples were collected with a face mask (Fig. 1) from both subjects at 30-min intervals for 6 hr following administration of the dose. All stool samples were collected quantitatively from both subjects for 3 days postdose.

Breath and stool samples were analyzed for $\delta^{13}C_{PDB}$ by gas isotope ratio mass spectrometry, as described above and in Chapter 10, this volume. Total CO_2 production for each subject was estimated from body surface area using a constant of 5 mmol/m²/min.

4. Results and Calculation of ^{13}C Recovery

$\delta^{13}C_{PDB}$ values for respiratory CO_2 are shown for both subjects in Fig. 5. Note that for both subjects, significant ^{13}C enrichment was obtained in the first postdose sample taken only 30 min after the dose was administered. The slightly more rapid rise in the ^{13}C content of the breath from subject 2 might suggest that the presence of pancreatic amylase in his small intestine permitted more rapid digestion, absorption, and oxidation of the labeled substrate. The slower rise in the ^{13}C content of breath from subject 1 may reflect a slightly longer transit time to a site where carbohydrate could be digested and absorbed; in the absence of pancreatic amylase, a likely site for this to occur is the colon, where bacteria ferment the carbohydrate to

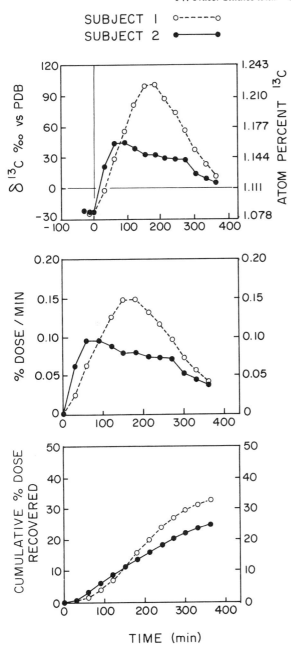

SUBJECT 1 o------o
SUBJECT 2 •——•

Fig. 5 Time course of ^{13}C excretion in respiratory CO_2 from subjects 1 and 2 following administration of ^{13}C starch. Figures show (from the top) δ^{13}C values of individual breath samples, % dose excreted/minute, and cumulative % dose recovered.

volatile fatty acids, which are absorbed readily and oxidized for energy production.

For subsequent mass calculations, all $\delta^{13}C_{PDB}$ values were converted to fractional abundances (F), as described in Table 2 (formulas a–c). Then, for each subject, a mean fractional abundance for all baseline breath samples was determined and used to calculate atom % excess (APE) for each post-dose breath sample (Table 2, formula d).

To determine the rate at which the dose was excreted and the proportion of the dose recovered in respiratory CO_2, formulas f through h (Table 2) were carried out in sequence for each postdose time point. The amount of excess ^{13}C excreted per minute was the product of the APE of the sample and the total CO_2 production rate (Table 2, formula f). Dividing this parameter by the amount of excess ^{13}C in the dose yielded the % dose excreted/min (Table 2, formula g). The cumulative % dose recovered (CUMPCD) was the area under the curve generated by plotting % dose/min versus time (Fig. 5). Any technique for integrating area under a curve will work, but a formula based on trapezoidal integration (Table 2, formula h) is offered here because of its simplicity and ease of calculation. Cumulative % dose recovered is shown for both subjects in Fig. 5. At 6 hr postdose, 32% of the dose was recovered in the breath of subject 1 and 25% of the dose was recovered in the breath of subject 2. Thus, subject 1 was apparently able to digest and oxidize a substantial proportion of the ^{13}C starch despite the probable absence of pancreatic amylase from her digestive system.

Recovery of ^{13}C in stool was determined from the equations in Table 3. The mass of carbon in each stool was calculated as the product of the dry weight of the stool and its percent carbon (formula a). The number of mmol of stool carbon is determined by dividing the mass of stool carbon by the

---

Table 3
Formulas Required for Determining Malabsorption of ^{13}C-Labeled Substrates in Stool

a. Mass of stool carbon (mg C) = stool dry weight $\times$ stool % carbon

b. $\begin{matrix} \text{mean atomic weight} \\ \text{of stool C} \end{matrix}$ = (12)(atom % ^{12}C in stool) + (13)(atom % ^{13}C in stool)

c. mmol stool C = $\dfrac{\text{mass of stool C (mg)}}{\text{mean atomic wt of stool C}}$

d. mmol excess ^{13}C in stool = (mmol stool C)($F_{postdose} - F_{baseline}$)

e. % dose malabsorbed = $\dfrac{\text{mmol excess } ^{13}C \text{ in stool}}{\text{mmol excess } ^{13}C \text{ in dose}} \times 100$

Table 4
Stable Carbon Isotope Composition and Recovery of Dose in Stool from Subjects 1 and 2

Subject	$\delta^{13}C_{PDB}$	F	Stool dry weight(g)	Stool % C	Excess ^{13}C (mmol)	% Dose recovered
Subject 1						
baseline mean	−24.5	0.010843	—	—	—	—
postdose 1	−12.8	0.010972	2.122	53.0	0.0120	0.7
postdose 2	−16.1	0.010935	2.962	52.5	0.0118	0.7
postdose 3	−22.6	0.010864	4.432	47.7	0.0035	0.2
Total					0.0273	1.6
Subject 2						
baseline mean	−22.0	0.010871	—	—	—	—
postdose 1	−16.5	0.010931	4.023	45.0	0.0090	0.4
postdose 2	−16.7	0.010929	3.875	44.9	0.0084	0.4
Total					0.0174	0.8

mean atomic weight of stool carbon (formulas b and c). The mmol of excess ^{13}C present in a postdose stool sample was the product of the mmol of total carbon in the sample and the difference between the fractional abundances of postdose and baseline stool samples (formula d). Finally, the percentage of the dose malabsorbed or recovered in each stool was obtained by dividing the mmol of excess ^{13}C present in each stool sample by the mmol of excess ^{13}C present in the dose (formula e).

Data for stools collected over 72 hr postdose from subject 1 and 2 are shown in Table 4. The mmol excess ^{13}C and the percentage of the dose recovered in each postdose stool sample was calculated using the formulas in Table 3. For both subjects, the amount of excess ^{13}C recovered in each stool sample and over the entire 72-hr collection period was extremely low and metabolically insignificant, indicating that both almost completely absorbed the ^{13}C-labeled rice starch. Even for adults, malabsorption of 1 to 5% of a carbohydrate meal is considered normal.

5. Discussion

Both subjects showed nearly complete absorption of the ^{13}C-labeled starch, as evidenced by the extremely low recovery of ^{13}C in the stool. This suggests that, despite the presumed absence of pancreatic amylase in the 3-month-old infant (subject 1), some alternative mechanism of starch digestion was operative and allowed her to assimilate the ^{13}C starch. Based on recovery of ^{13}C in respiratory CO_2, both subjects oxidized a significant proportion of the dose for energy production.

Determination of whole-body oxidation rates for ^{13}C-labeled nutrients is based on recovery of excess ^{13}C in respiratory CO_2. To assess accurately the proportion of the labeled substrate that was oxidized during the study, a correction must be applied to account for loss of labeled carbon that occurs during transit through the bicarbonate pools. Losses to the bicarbonate pools are evaluated by measuring ^{13}C recovery in respiratory CO_2 following intravenous administration of $NaH^{13}CO_3$. The percent of a labeled nutrient that has actually been oxidized by a human or animal subject is then calculated by dividing the percentage of the dose of the labeled nutrient that has been recovered in respiratory CO_2 by the fraction of an intravenous $NaH^{13}CO_3$ dose recovered in respiratory CO_2. For example, if 35% of a dose of ^{13}C glucose was recovered on one test day, and 85% of an intravenous dose of $NaH^{13}CO_3$ was recovered from the same subject on an alternate test day, then the true whole-body oxidation of ^{13}C-glucose was 35%/0.85 = 41%.

While some investigators elect to measure losses to the bicarbonate pool on every subject undergoing ^{13}C nutrient oxidation studies, a high degree of intraindividual relative to interindividual variation in bicarbonate kinetics suggests that this additional effort is not warranted (Irving *et al.*, 1983). This implies that average values for bicarbonate recovery could be developed for specific populations (e.g., normal adults, lactating women, premature infants) and applied to additional subjects in similar metabolic, nutritional, and pathological states.

Bicarbonate recovery from infants 2 to 5 months of age (Irving *et al.*, 1985) and from normal adults (Irving *et al.*, 1983) has been found to be approximately 50% over 4 hr, with no further recovery beyond that time. Therefore, true oxidation of the ^{13}C-labeled starch by subject 1 was 32%/0.5 = 64% of the dose, and by subject 2 was 25%/0.5 = 50% of the dose. These recoveries compare favorably with those documented for children of comparable age for starch (Shulman *et al.*, 1986) and simple carbohydrates (Murray *et al.*, 1990).

From these data, budgets can be constructed to illustrate the fate of the ^{13}C-labeled starch. For subject 1, 64% of the dose was oxidized for energy production and 1.6% of the dose was recovered in the stool, so approximately 34.4% of the dose must have gone to tissue synthesis or growth. For subject 2, 50% of the dose was oxidized and 0.8% of the dose was recovered in stool, leaving 49.2% of the dose for tissue synthesis. Knowing that virtually all of the dose was absorbed by both subjects, the cumulative percentage of the dose recovered in the breath from each subject suggests that nutrient partitioning may differ as a result of the age difference between these two subjects. While it is impossible to generalize from two subjects, it is interesting to speculate that the greater extent of substrate oxidation by subject 1 may reflect a higher energy demand for 3-month-old infants relative to 3-year-old children. However, it should be clear that, with adequate sample sizes, these techniques make it possible to test for differences in nutrient absorption and allocation between different populations.

IV. COMMENTS

The protocol described above was clearly designed to investigate a specific nutritional problem, that is, the utilization of dietary starch by young infants. However, the general design of this protocol is such that it answers several questions that might be asked with other labeled substrates, other study populations, or different species. These questions include: (1) How much of the nutrient is absorbed versus malabsorbed? (2) Of the proportion absorbed, how much is oxidized for energy production and how much is allocated for growth or tissue synthesis? and (3) What is the temporal aspect of nutrient absorption and oxidation? These are general questions that can be asked in a variety of contexts with different nutrients and different study populations. These questions might be directed toward studies of nutrient utilization during different developmental stages, in different metabolic states, or during recovery from disease or trauma.

The benefits of using stable rather than radioactive isotope tracers when working with human subjects are obvious. However, the benefits are less obvious for studies with large animals, and the higher costs of labeled substrates and analytical equipment required for stable versus radioactive tracer work have no doubt deterred investigators from using stable isotopes in animal studies. Since tracer doses are determined per unit of body weight, the cost of the tracer to conduct even a single stable isotope study with a cow or horse would be rather high.

However, alternative approaches for stable isotope studies with large animals exist. For example, carbon transfer to milk in dairy cows has been studied by introducing changes in the natural stable carbon isotope composition of the feed (Boutton *et al.,* 1988). Cows were switched from an alfalfa-based (^{13}C-depleted) to a corn-based (^{13}C-enriched) diet, and the rate of change of the carbon isotope composition of the milk was measured. These data were used to construct a compartmental model, which revealed several unexpected and important pieces of information regarding the nature of the metabolic precursor pool from which milk is synthesized. The point to be made here is that significant physiological information was obtained by simple manipulation of the diet without the use of highly enriched, expensive isotopic tracers. Many other questions related to understanding the fate of dietary nutrients in large animals could be addressed by manipulating the natural ^{13}C content of the diet. The potential of this methodology in metabolic studies on large animals remains largely unexplored, but may provide a viable alternative to the difficulties and hazards associated with radioisotope usage.

ACKNOWLEDGMENTS

This manuscript benefited from reviews by David D. Briske, Peter D. Klein, and Brian A. McGaw. The assistance of Sylvia Dudash with manuscript preparation is gratefully acknowledged. Support for the preparation of this manuscript was provided by project H-6945 of the Texas Agricultural Experiment Station.

REFERENCES

Ajami, A. M., and Watkins, J. B. (1983). Residual gas contaminants in sterilized serological tubes: A problem re-examined. *Clin. Chem.* **29**, 725–726.

Anonymous (1987). Availability of enriched stable isotope reference materials for medical and biological studies. *Am. J. Clin. Nutr.* **46**, 534.

Bier, D. M., and Matthews, D. E. (1982). Stable isotope tracer methods for *in vivo* investigations. *Fed. Proc.* **41**, 2679–2685.

Boutton, T. W., Wong, W. W., Hachey, D. L., Lee, L. S., Cabrera, M. P., and Klein, P. D. (1983). Comparison of quartz and pyrex tubes for combustion of organic samples for stable carbon isotope analysis. *Anal. Chem.* **55**, 1832–1833.

Boutton, T. W., Bollich, C. N., Webb, B. D., Sekely, S. L., and Klein, P. D. (1987). ^{13}C-labeled rice produced for dietary studies. *Am. J. Clin. Nutr.* **45**, 844.

Boutton, T. W., Tyrrell, H. F., Patterson, B. W., Varga, G. A., and Klein, P. D. (1988). Carbon kinetics of milk formation in Holstein cows in late lactation. *J. Anim. Sci.* **66**, 2636–2645.

Chevalier, R., Pelletier, G., and Gagnon, M. (1984). Sampling technique for collection of expired CO_2 in studies using naturally labeled ^{13}C in calves. *Can. J. Anim. Sci.* **64**, 495–498.

Flatt, W. P. (1969). Methods of calorimetry. B. Indirect. In "Nutrition of Animals of Agricultural Importance" (D. Cuthbertson, ed.), pp. 491–520. Pergamon Press, New York.

Hachey, D. L., Wong, W. W., Boutton, T. W., and Klein, P. D. (1987). Isotope ratio measurements in nutrition and biomedical research. *Mass Spectrom. Rev.* **6**, 289–328.

Halliday, D., and Rennie, M. J. (1982). The use of stable isotopes for diagnosis and clinical research. *Clin. Sci.* **63**, 485–496.

Haycock, G. B., Schwartz, G. J., and Wisotsky, D. H. (1978). Geometric method for measuring body surface area: a height-weight formula validated in infants, children, and adults. *J. Pediatr.* **93**, 62–66.

Irving, C. S., Wong, W. W., Shulman, R. J., Smith, E. O., and Klein, P. D. (1983). [^{13}C] bicarbonate kinetics in humans: Intra- vs. interindividual variations. *Am. J. Physiol.* **245**, R190–R202.

Irving, C. S., Lifschitz, C. H., Wong, W. W., Boutton, T. W., Nichols, B. L., and Klein, P. D. (1985). Characterization of HCO_3/CO_2 pool sizes and kinetics in infants. *Pediatr. Res.* **19**, 358–363.

Irving, C. S., Klein, P. D., Navratil, P. R., and Boutton, T. W. (1986). Measurement of $^{13}CO_2/^{12}CO_2$ abundance by nondispersive infrared heterodyne ratiometry as an alternative to gas isotope ratio mass spectrometry. *Anal. Chem.* **58**, 2172–2178.

Jones, P. J. H., Pencharz, P. B., Bell, L., and Clandinin, M. T. (1985). Model for determination of ${}^{13}C$ substrate oxidation rates in humans in the fed state. *Am. J. Clin. Nutr.* **41**, 1277–1282.

Klein, P. D., and Klein, E. R. (1985). Applications of stable isotopes to pediatric nutrition and gastroenterology: Measurement of nutrient absorption and digestion using ${}^{13}C$. *J. Pediatr. Gastroenterol. Nutr.* **4**, 9–19.

Klein, P. D., Schoeller, D. A., and Klein, E. R. (1981). Recent developments in biomedical applications of stable isotopes. In "Symposium on HPLC, GC, and MS. Applications in Clinical Research" (A. M. Lawson, ed.), pp. 119–134. Academic Press, New York.

Kollman, V. H., Hanners, J. L., Hutson, J. Y., Whaley, T. W., Ott, D. G., and Gregg, C. T. (1973). Large-scale photosynthetic production of carbon-13 labeled sugars: the tobacco leaf system. *Biochem. Biophys. Res. Comm.* **50**, 826–831.

Lefebvre, P. J. (1985). From plant physiology to human metabolic investigations. *Diabetologia* **28**, 255–263.

Matthews, D. E., and Bier, D. M. (1983). Stable isotope methods for nutritional investigations. *Ann. Rev. Nutr.* **3**, 309–339.

McKirnan, M. D., White, F. C., Guth, B. D., Longhurst, J. C., and Bloor, C. M. (1986). Validation of a respiratory mask for measuring gas exchange in exercising swine. *J. Appl. Physiol.* **61**, 1226–1229.

Metges, C., Kempe, K., and Schmidt, H. L. (1990). Dependence of the carbon isotope contents of breath carbon dioxide, milk, serum, and rumen fermentation products on the $\delta^{13}C$ value of food in dairy cows. *Br. J. Nutr.* **63**, 187–196.

Milne, E., and McGaw, B. A. (1987). The applicability of evacuated serological tubes for the collection of breath for isotopic analysis of CO_2 by isotope ratio mass spectrometry. *Biomed. Environm. Mass Spectrom.* **15**, 467–472.

Murray, R. D., Boutton, T. W., Klein, P. D., Gilbert, M., Paule, C. L., and MacLean, W. C. (1990). Comparative absorption of [${}^{13}C$] glucose and [${}^{13}C$] lactose by premature infants. *Am. J. Clin. Nutr.* **51**, 59–66.

Nakamura, K., Schoeller, D. A., Winkler, F. J., and Schmidt, H. L. (1982). Geographical variations in the carbon isotope composition of the diet and hair in contemporary man. *Biomed. Mass Spectrom.* **9**, 390–394.

Sauer, P. J., Van Aerde, J. E., Pencharz, P. B., Smith, J. M., and Swyer, P. R. (1986). Glucose oxidation rates in newborn infants measured with indirect calorimetry and [U-${}^{13}C$] glucose. *Clin. Sci.* **70**, 587–593.

Schoeller, D. A., and Klein, P. D. (1978). A simplified technique for collecting breath CO_2 for isotope ratio mass spectrometry. *Biomed. Mass Spectrom.* **5**, 29–31.

Schoeller, D. A., and Klein, P. D. (1979). A microprocessor controlled mass spectrometer for the fully automated purification and isotopic analysis of breath carbon dioxide. *Biomed. Mass Spectrom.* **6**, 350–355.

Schoeller, D. A., Schneider, J. F., Solomons, N. W., Watkins, J. B., and Klein, P. D. (1977). Clinical diagnosis with the stable isotope ${}^{13}C$ in CO_2 breath tests: methodology and fundamental considerations. *J. Lab. Clin. Med.* **90**, 412–421.

Schoeller, D. A., Klein, P. D., Watkins, J. B., Heim, T., and MacLean, W. C. (1980). ${}^{13}C$ abundances of nutrients and the effect of variations in ${}^{13}C$ isotopic abundances of test meals formulated for ${}^{13}CO_2$ breath tests. *Am. J. Clin. Nutr.* **33**, 2375–2385.

Schoeller, D. A., Klein, P. D., MacLean, W. C., Watkins, J. B., and van Santen, E. (1981). Fecal ${}^{13}C$ analysis for the detection and quantitation of intestinal malabsorption. *J. Lab. Clin. Med.* **97**, 439–448.

Schoeller, D. A., Brown, C., Nakamura, K., Nakagawa, A., Mazzeo, R. S., Brooks, G. A., and Budinger, T. F. (1984). Influence of metabolic fuel on the ${}^{13}C/{}^{12}C$ ratio of breath CO_2. *Biomed. Mass Spectrom.* **11**, 557–561.

Segal, K. R. (1987). Comparison of indirect calorimetric measurements of resting energy expenditure with a ventilated hood, face mask, and mouthpiece. *Am. J. Clin. Nutr.* **45,** 1420–1423.

Shreeve, W. W., Cerasi, E., and Luft, R. (1970). Metabolism of [2-^{14}C] pyruvate in normal, acromegalic, and HGH-treated human subjects. *Acta Endocrinol.* **65,** 155–169.

Shulman, R. J., Kerzner, B., Sloan, H. R., Boutton, T. W., Wong, W. W., Nichols, B. L., and Klein, P. D. (1986). Absorption and oxidation of glucose polymers of different lengths in young infants. *Pediatr. Res.* **20,** 740–743.

Svejcar, T. J., Boutton, T. W., and Trent, J. D. (1990). Assessment of carbon allocation with stable carbon isotope labeling. *Agron. J.* **82,** 18–21.

Tieszen, L. L., Boutton, T. W., Tesdahl, K. G., and Slade, N. A. (1983). Fractionation and turnover of stable carbon isotopes in animal tissues: Implications for δ^{13}C analysis of diet. *Oecologia* **57,** 32–37.

von Unruh, G. E., Hauber, D. J., Schoeller, D. A., and Hayes, J. M. (1974). Limits of detection of carbon-13 labeled drugs and their metabolites in human urine. *Biomed. Mass Spectrom.* **1,** 345–349.

Wilson, G. F., MacKenzie, D. D. S., Brookes, I. M., and Lyon, G. L. (1988). Importance of body tissues as sources of nutrients for milk synthesis in the cow, using ^{13}C as a marker. *Br. J. Nutr.* **60,** 605–617.

Winchell, H. S., Stahelin, H., Kusubov, N., Slanger, B., Fish, M., Pollycove, M., and Lawrence, J. (1970). Kinetics of CO_2-HCO_3 in normal adult males. *J. Nucl. Med.* **11,** 711–715.

Wolfe, R. R. (1984). "Tracers in Metabolic Research. Radioisotope and Stable Isotope/Mass Spectrometry Methods." Alan R. Liss, Inc., New York.

Wolfe, R. R., Shaw, J. H. F., Nadel, E. R., and Wolfe, M. H. (1984). Effect of substrate intake and physiological state on background $^{13}CO_2$ enrichment. *J. Appl. Physiol: Respirat. Environ. Exercise Physiol.* **56,** 230–234.

Uses and Procedures for ^{11}C

15

Intact Organism, Short-Term Studies Using ^{11}C

Richard D. Spence[1]
Peter J.H. Sharpe

Biosystems Research Group
Department of Industrial Engineering
Texas A&M University
College Station, Texas 77843

I. INTRODUCTION

Experimental investigation of biochemical pathways and biophysical mechanisms is difficult because in many cases more than one pathway or mechanism is involved. Ideally, a physiological tracer should be used to follow the uptake, transport, and assimilation of materials such as carbon and nitrogen to characterize the movements and mechanisms of physiological processes. Real-time measurements of net photosynthesis and dark respiration of plants have been possible since the development of the infrared gas analyzer (IRGA), allowing the intensive investigation of mechanisms and dynamics of CO_2 assimilation in green plants. Comparable research on the movement of carbon within the plant, however, requires another technique that allows real-time observations of carbon transport. This chapter describes how the short-lived radioisotope ^{11}C can be used to conduct plant physiological studies that are difficult or impossible to make using other isotopes such as ^{14}C.

1. Current address: Division of Science and Engineering, University of Texas of the Permian Basin, 4901 East University, Odessa, Texas 79762.

A. ^{14}C

The long-lived radioisotope ^{14}C has been used extensively in plant physiology studies for some years and from it much useful information has been obtained. Generally, however, it has not been possible to closely link CO_2 uptake and carbon movement using ^{14}C. The use of ^{14}C has several drawbacks (Strain *et al.*, 1983). The low-energy beta particles (β^-) (0.156 MeV) emitted by the decay of ^{14}C cannot escape the plant tissue. It is necessary to harvest tissues destructively and to quantify radioactivity by the scintillation counting of prepared samples. Carbon translocation must thus be determined by time-series and statistical procedures. Physiological variability requires many replications of each measurement, raising the cost of such experiments in terms of time, manpower, material, and effort. Due to the long half-life of ^{14}C (5770 years), the isotope, once absorbed, persists in the tissue, preventing both repeated, long-term measurements of an exposed individual and real-time measurements of carbon transport.

B. ^{11}C

Carbon-11 has several advantages as a tracer over ^{14}C (Fares *et al.*, 1978): (1) ^{11}C, like most short-lived radioisotopes, decays by positron (β^+) emission followed by positron-electron annihilation, emitting two γ rays at 180° to each other with sufficient energy (0.511 MeV) to be detected through several centimeters of tissue *in vivo*; (2) the short half-life (20.4 min) of ^{11}C means that the isotope is effectively eliminated from the tissue after only a few hours, allowing multiple, sequential experiments on the same plant under the same or contrasting sets of environmental conditions; (3) the nondestructive nature of the isotope means that the plant can serve as its own control, reducing the need for and cost of multiple replications; and (4) the half-life is comparable to the turnover times of the photosynthetic pool and velocity of transport, allowing dynamic measurements that cannot be done with long-lived radioisotopes.

In experimental protocols, ^{11}C is usually introduced to the plant in the form of $^{11}CO_2$, which is fixed during ordinary photosynthesis. The movement of labeled photosynthate is monitored by detectors placed along the petiole, stem, roots, and even other leaves. In conjunction with gas-exchange analyzers, a number of simultaneous, nondestructive measurements are made routinely using this technique. These include: (1) transpiration rate; (2) stomatal conductance; (3) net photosynthesis by ^{12}C exchange rate and ^{11}C assimilation rate; (4) net rate of photosynthate storage in the leaf; (5) net rate of photosynthate export from the leaf; (6) turnover time of exportable products; (7) pool size of exportable products; (8) axial speed of translocation; (9) activity level of translocated photosynthates; and (10) unloading rate into any given sink tissue.

The scope and breadth of these measurements means that: (1) the systems behavior of the overall plant physiological response may be characterized; (2) dynamic carbon allocation patterns may be observed; (3) genetically and environmentally induced changes in plant physiology may be evaluated; and (4) genetic and environmental effects on a quantity of interest may be estimated. A host of environmental and biological factors can be tested with the ^{11}C technique. The technique thus can lead to improved qualitative and quantitative understanding of plant responses to environmental stimuli. In fact, this technique has already elucidated many aspects of plant physiology and plant responses to various environmental factors.

The first ^{11}C studies on plant physiology were conducted in the late 1930s by a group at the University of California, Berkeley, headed by Ruben (Ruben *et al.*, 1939). Carbon-11 studies were revived by Moorby in England in the early 1960s (Moorby *et al.*, 1963). Beginning in the 1970s, ^{11}C work has been carried out by Fensom, a Canadian working in Scotland (Fensom *et al.*, 1977), and by Troughton and Minchin in New Zealand (Troughton *et al.*, 1974; Minchin and Troughton, 1980). In the United States, Goeschl, Fares, and Magnuson have been conducting ^{11}C studies since the middle 1970s, first as part of the Biosystems Research Group at Texas A&M University (Fares *et al.*, 1978), and currently at the Phytotron at Duke University (Goeschl *et al.*, 1988). Unless otherwise indicated, the procedures for the production and use of ^{11}C described in this chapter are based upon those currently used at Duke.

II. [11]C PRODUCTION AND USE

A. Manufacture of [11]CO_2

The ^{11}C isotope can be produced using one of two approaches. In *batch production*, all of the isotope necessary to perform the experiment is produced in a short period of time, immediately before the experiment begins. Usually storage and transport from the reactor to the plant research facility are involved. The short half-life of ^{11}C requires that far more [11]CO_2 be produced by batch production than will actually be used. In *continuous production*, the [11]CO_2 is produced concurrently with the experiment and is used as it is produced. The method used depends on reactor power, availability, and proximity to the plant research facility. Most ^{11}C work has used the batch production method. At Texas A&M University in 1976 and at Duke University in the early 1980s, however, [11]CO_2 was produced continuously during experiments (Fares *et al.*, 1978, 1983).

There are several reactions by which ^{11}C can be produced from ^{12}C or from other elements, including [10]B(d,n)[11]C (Ruben *et al.*, 1939; Moorby *et al.*, 1963; More and Troughton, 1972), $^{12}C(\gamma,n)^{11}C$ (Fensom *et al.*, 1977), $^{12}C(p,d)^{11}C$, $^{12}C(^3He, ^4He)^{11}C$, and $^{12}C(d,t)^{11}C$ (see Fares *et al.*, 1978; Goeschl *et al.*, 1988). Currently at Duke University, $^{11}CO_2$ is batch produced by a CS-30 cyclotron, located on the ground floor of the Duke University Hospital, using the reaction $^{14}N(p,\alpha)^{11}C$ (McKinney *et al.*, 1988). The target contains at first 99.999% nitrogen gas (N_2), which forms ^{11}C upon bombardment with protons. The active carbon forms $^{11}CO_2$ with the residual O_2 in the target gas. Following irradiation, the $^{11}CO_2$ is transferred to a storage tank (McKinney *et al.*, 1988).

Production provides a single batch of $^{11}CO_2$ from a short irradiation consisting of $5\mu A$ of beam for 6–8 min. A typical tracer experiment requires approximately two hours of continuous $^{11}CO_2$ supplied to the plant to reach isotopic steady state. The amount of activities in Becquerels, A_{batch}, that must be produced in order to supply the plant with K Bq/min and an activity of $A(t)$ Bq remaining at the end of an experiment of length t in minutes is (McKinney *et al.*, 1988):

$$A_{batch} = [A(t) + K/\lambda]e^{\lambda t} - K/\lambda \qquad (1)$$

Assuming that $^{11}CO_2$ is supplied to the plant at a flow rate of 1000 cm³/min at an activity level of 3700 Bq/cm³, then $K = 3.7 \times 10^7$ Bq/min. Allowing for ten times the required amount of activity at the end of the experiment, then $A(t) = 3.7 \times 10^7$ Bq. For a typical experiment of length $t = 120$ min, and using $t_{1/2} = 20.4$ min for ^{11}C, solving for A_{batch} yields 8500 Bq (equivalent to about 230 nCi). It can thus be seen that extremely low activity levels of ^{11}C are sufficient for most physiological characterizations. This low activity level renders ^{11}C harmless both to the plant and to the experimenter.

Air containing $^{12}CO_2$ at a desired concentration is mixed with the $^{11}CO_2$ through a flow controller to provide a constant gas flow. As the air/$^{12}CO_2$/$^{11}CO_2$ mixture passes through the plant leaf cuvette (see below) its radioactivity is monitored continuously by a scintillation detector connected to a ratemeter. The output of the ratemeter drives a set point controller which, in turn, adjusts a flow controller on the $^{11}CO_2$ storage tank to maintain a preset constant ^{11}C activity in the cuvette. The $^{11}CO_2$ makes up a very small fraction of the gas mixture so that a constant flow is assured regardless of the status of the $^{11}CO_2$ flow (McKinney *et al.*, 1988).

B. Pulse and Steady-State Labeling

There are two ways of introducing the labeled material to the system: by a *pulse* of very short duration, usually a few minutes, or in a continuous stream until isotopic *steady-state* is achieved. The techniques emphasize different physiological processes, but more often than not the decision for pulse or steady-state labeling is based upon convenience, costs, instrumentation capabilities, and personal preferences. Virtually all of the earliest uses of ^{11}C involved pulse labeling, and many groups, for example, Minchin's in New Zealand, still do so. The first steady-state labeling of $^{11}CO_2$ was done by the Biosystems group at Texas A&M in the 1970s (Fares *et al.*, 1978).

Pulse labeling may be done without much of the effort and expensive instrumentation required by steady-state labeling systems. Pulse labeling in plant physiological studies is generally sufficient for investigating loading rates, phloem transport speeds, and approximate allocation patterns. In pulse labeling the total amount of radioisotope delivered must be measured, but generally it is not necessary to know the specific activity, that is, the ratio of labeled compound to the total amount of the subject compound (Geiger, 1980).

Some physiological processes, however, cannot be investigated completely under pulse labeling. Not all compounds to be labeled reach isotopic steady-state simultaneously. This is due to differences in turnover times between various pools of intermediates. Pulse labeling aggravates this disparity by favoring the labeling of small pools with short turnover times. For example, newly formed starch in photosynthesizing leaves reaches isotopic saturation in a few minutes, while sucrose requires over 90 minutes to reach saturation. As a result, pulse labeling emphasizes the relative contribution of starch over sucrose to the photosynthate pool (Geiger, 1980).

Steady-state labeling, also called "extended square wave" or "step input" labeling (Goeschl *et al.*, 1988), avoids this bias by bringing all primary pools to isotopic saturation, or steady-state, so that the specific activity of the pool is commensurate to its relative size and turnover time. Analysis of subsequent movement between pools is facilitated. For example, a pool accumulates isotope at a constant rate if the pool supplying the isotope has reached isotopic steady-state or saturation. Thus pool sizes and rates of export can be calculated easily, something hard to do using pulse labeling (Geiger, 1980; Goeschl *et al.*, 1988).

Once isotopic steady-state in the leaf is achieved (Fig. 1), the $^{11}CO_2$ tracer input can be stopped. From the decreasing levels of ^{11}C in the leaf, this phase, lasting some 120 minutes after isotope labeling is stopped, is referred to as the "washout curve." The shape and duration of the washout curve for each pool can be analyzed for pool size and turnover time (Jacquez, 1985).

C. Plant Chamber

Figure 2 provides a general illustration of the $^{11}CO_2$ testing facility, showing the plant in the chamber, detectors, and gas delivery and monitoring instrumentation. The plant of choice, grown under the environmental conditions chosen for a particular study, is moved into the controlled environment chamber (CEC) in which the $^{11}CO_2$ characterization is made, 24–48 hr

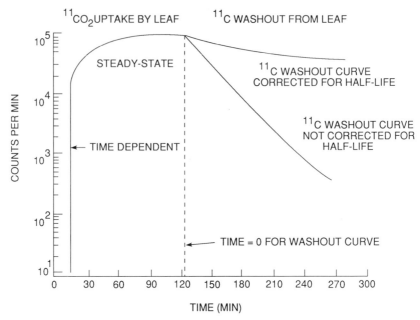

Fig. 1 A typical representation of ¹¹C uptake in a leaf in real time during steady-state labeling. Adapted from Goeschl *et al.* (1988).

before the start of the experiment. The CEC measures about 1 m in height by 1 m in width by 0.75 m in depth, is equipped with lighting, temperature controls, and a humidifier. The lighting source is a high-intensity discharge (HID) halide lamp capable of providing $750 - 850 \ \mu E/m^2/sec$ illumination at the level of the canopy. Photoperiod is regulated by a timer. Chamber temperature during the photoperiod is maintained at $29 \degree C$ by circulating-air fans. The humidifier remains at 70% relative humidity.

A portion of one of the plant leaves is enclosed with a clear plastic water-cooled cuvette and supplied with a continuous flow of air at a given steady-state concentration of $^{12}CO_2$ (Fig. 2). Net photosynthesis, transpiration, and other physiologically related functions are monitored continuously. The detectors are arranged on the plant at the desired positions. The geometry of the leaves and the petiole of the leaf on which the experiment is conducted are measured and entered into the computer. On the morning of the experiment, as soon as the lights of the CEC come on, the background activity at each detector is determined and the detectors are calibrated with a standard ^{22}Na source. This information is automatically entered and stored in the computer.

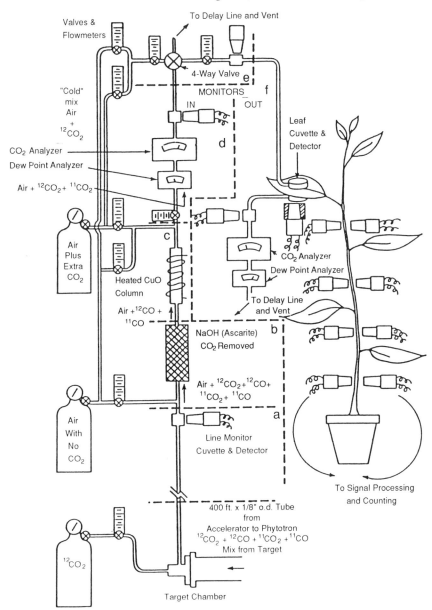

Fig. 2 The ¹¹CO₂ test procedure. To the right are the plant with the cuvette attached to the leaf and detectors placed at desired locations along the stem. An air mixture containing ¹¹CO₂ from the reactor (lower left) is monitored (upper left) for continuous activity before being fed to the plant. Adapted from Goeschl *et al.* (1988).

D. Detectors

Carbon-11 assimilation, translocation, and allocation throughout the plant are monitored by means of detector pairs placed at desired locations of study along the stems, leaves, and roots of the plants (see Fares *et al.*, 1978; 1983). The front end of the detector, facing the source of activity in the plant, can be fitted with a lead collar to give a desired window. Detectors are positioned in pairs on either side of the plant stem above and below the test leaf (Fig. 2). The detector pairs are operated in time coincidence at 180° with respect to each other. The distances between pairs and their positions with respect to the stem and with each other are carefully assigned to monitor carbon flow at precise locations.

E. Signal Processing and Counting

Standard electronics process the signals from the detectors, interfacing with the data processor for data acquisition. A controller serves as the "traffic director" for the many systems signals. The processor for this facility is a minicomputer. The processor (1) communicates with the controller system, sending commands to and receiving responses from the controller; (2) implements routines such as calibration and background corrections; (3) implements on-line data acquisition, reduction, storage, and display; (4) tests the on-line acquired data; and (5) implements utility programs. The minicomputer has graphic capabilities for visual display, and coordinates two magnetic tape drives, floppy disk drives, thermal printer, and multipen multicolor recorders (Fares *et al.*, 1978; Goeschl *et al.*, 1988).

F. ^{11}C Characterization

When all is ready the air/$^{12}CO_2$/$^{11}CO_2$ mixture is diverted to the cuvette at a constant ^{11}C activity. The leaf begins to take up the $^{11}CO_2$ through ordinary photosynthesis (Fig. 3). The buildup of ^{11}C activity in the leaf and its movement in the transport stream are continuously measured and updated by the computer. All metabolic pools in the leaf reach isotopic steady-state within 90–120 min. While the labeled mixture is supplied to the leaf, the computer analyzes the data in real time, printing out and graphing various activity and plant parameter data at 1-min intervals. Carbon-11 activity levels in the plant continue to be monitored after the $^{11}CO_2$ tracer input to the leaf is stopped ($^{12}CO_2$ and air continue to be supplied through the cuvette).

The activity at any given point along the plant (corrected for ^{11}C decay and the initial specific activity) is proportional to the concentration of carbon in the mobile pools, the transport system, and/or the rate of assimilation into sinks. Both mobile and stored activities generate highly characteristic patterns at a given position on the plant.

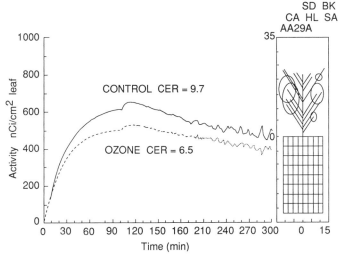

Fig. 3 Real-time display of ^{11}C activity in the needle of loblolly pines, one of which was fumigated with ozone according to the procedure of Spence *et al.* (1990).

Two hours after stopping the tracer input in the leaf (Fig. 3), only about 3% of the activity is left in the plant. The same experiment (or other experiments) could be repeated on the same plant under the same environmental conditions (Goeschl *et al.*, 1988). Thus the same plant can be used again and again, serving as its own control.

III. EXPERIMENTAL RESULTS OBTAINED BY THE ^{11}C TECHNIQUE

The first use of ^{11}C in plants was to measure carbon assimilation (Ruben *et al.*, 1939). Since then, ^{11}C has been used to study a wide variety of plant physiological processes related to carbon translocation.

A. Translocation

The pioneering ^{11}C work of Moorby and co-workers (beginning with Moorby *et al.*, 1963) provided the first *in vivo*, nondestructive measurements of photosynthate translocation in the phloem. Over the next decade, ^{11}C was used to measure rates of translocation and phloem velocities of several representative species. The data obtained have been used to formulate new models of the phloem transport process (e.g., Minchin, 1978; Minchin and

Troughton, 1980) or to test existing models such as the Münch–Horwitz osmotic pressure flow hypothesis (e.g., Goeschl and Magnuson, 1986; Magnuson *et al.*, 1986). Along the way a fuller understanding has been obtained of sieve tube physiology and function, loading and unloading rates and processes, and photosynthate concentrations and velocities in the sieve tubes.

B. Responses to Temperature

Early studies using traditional methods indicated that cold temperatures reduce stomatal conductance, net photosynthesis, and photosynthate translocation. Studies in the last decade using ^{11}C have elucidated some of the mechanisms that determine plant response to low temperature. Translocation in regions near a cold block on stems is slowed immediately; translocation in regions far from the block are also altered. The response to low temperature has been shown to be a whole-plant phenomenon and of considerable complexity (Pickard *et al.*, 1978a). Plants adapted to warm climates are more subject to chilling than plants of the same species adapted to cool climates (Potvin *et al.*, 1985).

C. Water Status

The osmotic driving mechanism of phloem transport indicates that it should be extremely sensitive to the water status of the plant; ^{11}C studies have confirmed this idea. Short-term, severe water stress stops translocation at about the same time as wilting becomes visible, while recovery of translocation following restoration of well-watered conditions is generally slow (Pickard *et al.*, 1978b). More gradual, moderate stress reveals a very interesting sequence of responses. In the initial stages of drought stress there is a reduction in export rate of photosynthate from the leaves and an increase in the concentration of exportable products and rate of storage within the leaf. As drought stress continues there are noticeable decreases in carbon exchange rate, transpiration, photosynthate export rate from the leaf, and in the pool size of exportable products (Goeschl *et al.*, 1988). This temporal sequence of responses to water stress, with a decrease in carbon export preceding a decrease in gas exchange, supports predictions made by models of phloem transport regarding the maintenance of phloem turgor (DeMichele *et al.*, 1978).

D. Agitation

In most ^{11}C studies it is rarely possible to grow plants or expose them to desired environmental conditions over a long term while they are simultaneously attached to the entire ^{11}C-testing apparatus. Movement from an experimental chamber to the ^{11}C chamber is thus necessary, and while physical trauma is minimized as much as possible, some agitation and flexing of the plant is unavoidable. Carbon-11 studies have shown that this mechanical agitation results in a general but short-term shutdown of the transport process. Recovery is usually complete, but varies over time from a few hours to two days depending on species and the individual plants (Goeschl *et al.*, 1988). More curious, recent studies have shown that at unpredictable times spontaneous blockage of phloem transport occurs in some species, for example, in cotton (Goeschl *et al.*, 1984) and loblolly pine (Sharpe *et al.*, 1989), for reasons not yet fully understood.

E. Atmospheric Pollutants

Recent studies using ^{11}C have provided compelling evidence that some of the most severe effects of atmospheric pollutants occur on the phloem transport system. Carbon allocation and thus growth processes may be altered long before visible damage is seen. Phloem loading was shown to be severely reduced when sulfur dioxide (SO_2) is applied to leaves of a C_3 plant, although no effect was seen immediately in C_4 plants, nor was there a change in transport velocity of the phloem (Minchin and Gould, 1986).

Characterizations of loblolly pine physiology using ^{11}C showed that moderate levels of ozone reduced stomatal conductance, carbon exchange rate, transpiration, photosynthesis, phloem velocity, photosynthate concentration in the phloem, and thus total carbon transport. In contrast to studies using SO_2 as a pollutant, ozone does not appear to inhibit the mechanism for phloem loading. Photosynthate is loaded into the phloem but allocation patterns to the roots are altered, so that photosynthate instead accumulates in the stems (Spence *et al.*, 1990). Ozone was thus shown to have the greatest impact on phloem transport processes, with a change in allocation of photosynthate from roots to stems.

IV. DISCUSSION

Despite the findings made through the use of ^{11}C, the technique has had its detractors. Some of the earliest of these were Sestak *et al.* (1971), who speculated, before the explosion of ^{11}C work later in the decade, that ^{11}C would be "of limited value on account of its short half-life of about 20 minutes." (p. 276) The reality is, however, that its short half-life is superbly suited for physiological studies.

Still, a certain defensiveness has inevitably crept into the writings of much ^{11}C work. It is interesting that almost every researcher from Ruben to the present has made the point that the reported work using ^{11}C could not have been completed as quickly, easily, or thoroughly using more conventional techniques.

In a more practical vein, the successful implementation of the ^{11}C technology is neither easy nor inexpensive. Sophisticated instrumentation and computers are required, plus specialized researchers and technicians to operate them. Specifically, there is the need for: (1) a dedicated accelerator near an environmentally controlled plant growth facility; (2) sophisticated electronics for tracer profile measurement; (3) an on-line computing facility for data acquisition and for the analysis of large data blocks in a relatively short time; (4) advanced mathematical methods of data analysis and dynamic modeling; and (5) a multidisciplinary approach coupled with a multidisciplinary team of scientists (Fares *et al.*, 1983).

The major practical disadvantage to the use of ^{11}C is that, at present, it is of limited use in field studies. So far, all studies of plant physiology or metabolism using ^{11}C have been laboratory studies. The difficulties in implementing ^{11}C in the field are obvious, revolving largely around the bulk and delicacy of the instrumentation and the need for a considerable source of power. Many of these limitations are not insurmountable, however, and limited field studies using ^{11}C could conceivably be implemented in the very near future.

The feasibility of bringing ^{11}C technology to the field has been enhanced by recent developments in storing high-activity $^{11}CO_2$ immediately after production (J. D. Goeschl and C. J. McKinney, personal communication). A small, portable, lead-lined pressurized storage container (whimsically called a "pig") has been charged with enough $^{11}CO_2$ so that even after several hours enough activity remains to conduct a series of viable experiments at another location. The pig is currently being tested at remote laboratory locations at Duke University. The next step entails loading detector and gas-monitoring instrumentation onto a truck with a pig and driving to an experimental field site. At the field site the gas-delivery cuvettes could be attached to a leaf on a tree as easily as to one in the laboratory. Thus the physiological responses of a plant to air pollutants or other environmental factors could be characterized for field conditions.

The benefit that balances the cost of establishing and maintaining sophisticated facilities, either for laboratory or future field studies, is the enormous amount of field data that can be collected simultaneously over short time intervals. Many of these parameters cannot be measured through conventional laboratory or field experimental techniques and those that can usually require months of painstaking replication. When the amount of information that can be obtained in such a short period from ^{11}C is compared to traditional experimental protocols, the technique can be seen to be remarkably cost-effective.

ACKNOWLEDGMENTS

Our ^{11}C studies were initiated as part of National Science Foundation Grant DEB 77-14408. These studies were extended as part of NSF BSR-86-14911. The carbon-11 studies of ozone effects on pines were funded by grants from the Air Quality/Forest Health Program, National Council of the Paper Industry for Air and Stream Improvement. For scientific collaboration and support, we thank Alan A. Lucier of NCASI; John D. Goeschl, Robert L. Musser, Charles E. Magnuson, and Collin J. McKinney, Biosystems Technology, Inc., Durham, North Carolina; Boyd R. Strain and the staff of the Duke University Phytotron; Walter W. Heck and Robert Philbeck of the Air Pollution Lab, USDA-ARS, North Carolina State University; and William E. Winner and Richard H. Waring, Oregon State University.

REFERENCES

DeMichele, D. W., Sharpe, P. J. H., and Goeschl, J. D. (1978). Towards the engineering of photosynthetic productivity. *Crit Rev in Bioeng* **3**, 23–91.

Fares, Y., DeMichele, D. W., Goeschl, J. D., and Baltuskonis, D. A. (1978). Continuously produced, high specific activity ^{11}C for studies of photosynthesis, transport, and metabolism. *Int J of Appl Radiat and Isot* **29**, 431–441.

Fares, Y., Goeschl, J. D., Magnuson, C. E., Nelson, C. E., Strain, B. R., Jaeger, C. H., and Bilpuch, E. J. (1983). A system for studying carbon allocation in plants using ^{11}C-labeled carbon dioxide. *Radiocarbon* **25**, 429–439.

Fensom, D. S., Williams, E. J., Aikman, D. P., Dale, J. E., Scobie, J., Ledingham, K. W. D., Drinkwater, A., and Moorby, J. (1977). Translocation of $^{11}CO_2$ from leaves of *Helianthus:* preliminary results. *Can J of Bot* **55**, 1787–1793.

Geiger, D. R. (1980). Measurement of translocation. In "Methods in Enzymology. Volume 69: Photosynthesis and Nitrogen Fixation" (A. San Pietro, ed.), pp. 561–571. Academic Press, New York.

Goeschl, J. D., and Magnuson, C. E. (1986). Physiological implications of the Münch–Horowitz theory of phloem transport: effects of loading rates. *Plant, Cell and Environ* **9**, 95–102.

Goeschl, J. D., Magnuson, C. E., Fares, Y., Jaeger, C. H., Nelson, C. E., and Strain, B. R. (1984). Spontaneous and induced blocking and unblocking of phloem transport. *Plant, Cell and Environ* **7**, 607–613.

Goeschl, J. D., Fares, Y., Magnuson, C. E., Scheld, H. W., Strain, B. R., Jaeger, C. H., and Nelson, C. E. (1988). Short-lived isotope kinetics: a window to the inside. In "Beltsville Symposia in Agricultural Research No. 11. Research Instrumentation for the 21st Century" (G. R. Beacher, ed.), pp. 21–53. Martinus Nijhoff Publishers, Dordrecht, The Netherlands.

Jacquez, J. A. (1985). "Compartmental Analysis in Biology and Medicine." Second edition. The University of Michigan Press, Ann Arbor.

Magnuson, C. E., Goeschl, J. D., and Fares, Y. (1986). Experimental tests of the Münch–Horowitz theory of phloem transport: Effects of loading rates. *Plant, Cell and Environ* **9**, 103–109.

McKinney, C. J., Fares, Y., Magnuson, C. E., Jaeger, C. H., Goeschl, J. D., and Need, J. L. (1988). Automatic system for the control of batch-produced $^{11}CO_2$ for continuous labeling experiments. *Rev of Sci Instrum* **59**, 467–469.

Minchin, P. E. H. (1978). Analysis of tracer profiles with applications to phloem transport. *J of Exp Bot* **29**, 1441–1450.

Minchin, P. E. H., and Gould, R. P. (1986). Effect of SO₂ on phloem loading. *Plant Sci* **43**, 179–183.

Minchin, P. E. H., and Troughton, J. H. (1980). Quantitative interpretation of phloem transport data. *Annu Rev of Plant Physiol* **31**, 191–215.

Moorby, J., Ebert, M., and Evans, N. T. S. (1963). The translocation of ¹¹C-labeled photosynthate in the soybean. *J of Exp Bot* **14**, 210–220.

More, R. D., and Troughton, J. H. (1973). Production of ¹¹CO₂ for use in plant translocation studies. *Photosynthetica* **7**, 271–274.

Pickard, W. F., Minchin, P. E. H., and Troughton, J. H. (1978a). Real time studies of carbon-11 translocation in moonflower I. The effects of cold blocks. *J of Exp Bot* **29**, 993–1001.

Pickard, W. F., Minchin, P. E. H., and Troughton, J. H. (1978b). Real time studies of carbon-11 translocation in moonflower II. The effects of metabolic and photosynthetic activity and of water stress. *J of Exp Bot* **29**, 1003–1009.

Potvin, C., Strain, B. R., and Goeschl, J. D. (1985). Low night temperature effect on C₄ grass species: II. Effect on photosynthate translocation. *Oecologia* **67**, 305–309.

Ruben, S., Hassid, W. Z., and Kamen, M. D. (1939). Radioactive carbon in the study of photosynthesis. *J of the Am Chem Soc* **61**, 661–663.

Sestak, Z., Catsky, J., and Jarvis, P. G. (1971). "Plant Photosynthesis Production: Manual of Methods." Dr. W. Junk N. V. Publishers, The Hague.

Sharpe, P. J. H., Spence, R. D., and Rykiel, E. J. (1989). Diagnosis of sequential ozone effects on carbon assimilation, translocation and allocation in cottonwood and loblolly pine. *NCASI Tech Bull,* No. 565, April 1989.

Spence, R. D., Sharpe, P. J. H., and Rykiel, E. J. (1990). Ozone alters carbon allocation in loblolly pine: Assessment with carbon-11 labeling. *Environ Pollut* **64**, 93–106.

Strain, B. R., Goeschl, J. D., Jaeger, C. H., Fares, Y., Magnuson, C. E., and Nelson, C. E. (1983). Measurement of carbon fixation and allocation using ¹¹C-labeled carbon dioxide. *Radiocarbon* **25**, 441–446.

Troughton, J. H., Moorby, J. and Currie, B. G. (1974). Investigations of carbon transport in plants: I. The use of carbon-11 to estimate various parameters of the translocation process. *J. of Exp. Bot.* **25**, 684–694.

Index

Accelerator mass spectrometer, in carbon
 dating measurements, 140–141
Acetyl-CoA carboxylase, herbicidal inhibition of, 105–107
   chemicals and equipment in study of,
      106
   imaging proportional counter analysis
      in, 106
   procedures used in study of, 106–107
Animal(s)
   ecosystems, calculations for determining
      $^{13}C/^{12}C$ ratios in, 215–216
   metabolism, ^{13}C labeled substrates in
      studies of, 219–220
   stable isotope studies with large, ap-
      proaches to, 239
Animal organ(s), $^{13}C/^{12}C$ ratios in different,
   208
Aquatic toxicology, degradation of organic
   xenobiotics and, 109–124. *See also*
   Xenobiotic(s), aquatic
   microcosm work in, 120–122
   radioactive tracers as research tools in,
      110
Aryloxyphenoxypropionic acid, herbicide,
   inhibition of acetyl-CoA carboxylase
   using, 105–107
Atmospheric air, sampling, $^{13}C/^{12}C$ fractionation in terrestrial plants and, 193–200
Atmospheric environment
   $\delta^{13}C$ values for CO_2, 174
   methane, 175–176
   stable carbon isotope ratios of, 174–176
Atmospheric pressure, control of, labeling
   chambers and, 17–18
Autoradiogram(s), whole plant, for aquatic
   translocation studies, 72

Autoradiography, aquatic translocation localization by, 71–72

Becquerel (Bq), defined, 3
Benthic system(s), polycyclic aromatic hydrocarbon metabolism in, microcosm
   study of, 121
Benzanthracene, aquatic degradation of
   lipid extraction scheme of metabolites
      after, 116
   metabolites formed from, 113
Bicarbonate(s), recovery of infant, ^{13}C labeled starch utilization studies and, 238
BOD bottles, phytoplankton photosynthesis
   using, 57
Bomb ^{14}C. *See* Bomb carbon
Bomb carbon, 147–151
   biological materials for, collection and
      treatment of, 149
   enrichment techniques for
      general and specific requirements of,
         148
      steady state diffusion model of, 148
      suitability of materials for, 148
   measurement of, techniques for, 149
   results, computations of, 149–150
   studies, usefulness of, 147
Bone(s),
   carbon dating of, 131–132
      minimal amount of sample required for,
         134
   collagen, extraction of, carbon dating
      measurements and, 136–137
   decontamination of, carbon dating and,
      132

259